Seawater Reverse Osmosis (SWRO) Desalination

Seawater Reverse Osmosis (SWRO) Desalination

Energy Consumption in Plants, Advanced Low-Energy Technologies, and Future Developments for Improving Energy Efficiency

Edited by

Seungkwan Hong, Kiho Park, Jungbin Kim, Abayomi Babatunde Alayande and Youngjin Kim

IWA PUBLISHING

Published by **IWA Publishing**
Unit 104–105, Export Building
1 Clove Crescent
London E14 2BA, UK
Telephone: +44 (0)20 7654 5500
Fax: +44 (0)20 7654 5555
Email: publications@iwap.co.uk
Web: www.iwapublishing.com

First published 2023

British Library Cataloguing in Publication Data
A CIP catalogue record for this book is available from the British Library

ISBN: 9781789061208 (Paperback)
ISBN: 9781789061215 (eBook)
ISBN: 9781789061222 (ePub)

This eBook was made Open Access in February 2023.

Doi: 10.2166/9781789061215

Contents

doi: 10.2166/9781789061215_0001

Chapter 1
Reverse osmosis

1.1 INTRODUCTION

Fresh water is essential for human life [1, 2]. Securing fresh water is one of the most important issues for humankind. However, human population has increased steadily, and millions of people are concentrated in metropolitan cities [3, 4]. The amount of natural freshwater is insufficient to meet the demands of the population [5]. Thus, fresh water must be produced from saline water resources to extend the capacity of the water supply. Desalination is defined as a methodology that removes/separates minerals and ions from saline water to obtain fresh water [6]. Therefore, the type and amount of dissolved minerals and ions in the feed water are very important for determining the appropriate desalination methodology. For example, seawater contains many minerals and ions that are very minute in size. Therefore, it is not sufficient to separate seawater to produce freshwater by using a simple filtration step such as microfiltration (MF) or ultrafiltration (UF). Thus, the characteristics of each desalination methodology should be considered to determine the appropriate method for each feed water type.

Desalination is usually conducted using two methodologies for separation: thermal-based desalination and membrane-based desalination. Before advances in membrane desalination technology, thermal-based desalination was widely utilized because it has many advantages, such as easy adaptability and producibility of high-quality freshwater. Humans have been supplying freshwater from saline water using thermal-based desalination for thousands of years. To satisfy the production capacity with increasing human population, this traditional method has been developed for advanced thermal desalination technologies such as multistage flash (MSF) and multi-effect distillation

(MED) [7]. Over the past several decades, with the development of membrane technology [8], membrane-based desalination has attracted the attention of many researchers due to its many advantages, such as easy scale-up and low energy consumption [9]. Many membrane-based desalination technologies, such as forward osmosis (FO), reverse osmosis (RO), capacitive deionization (CDI), and electrodialysis (ED), have been developed [10]. Among these membrane-based desalination technologies, RO has been regarded as one of the most widely utilized desalination technologies for freshwater production [11–14]. The increased preference for RO is obvious for seawater desalination. It can be clearly identified that the capacities of installed RO plants for seawater desalination were much higher than 65% of the overall installed capacities for seawater desalination in 2013 [15]. The overall price of desalination systems has been reduced. Therefore, it is expected that desalination capacity will continuously increase [9]. Under these circumstances, RO systems are a major technology for the utilization of seawater desalination.

Low energy consumption is the most attractive advantage of RO systems for seawater desalination [16, 17]. The main reason RO systems consume lower energy than thermal-based desalination is that phase transition is not required for the separation from feed water to permeate water (freshwater) and concentrated brine. Meanwhile, the energy consumption and the required operating pressure in the RO system are directly correlated with the feed water concentration [11, 12]. Therefore, it is important to consider the feed concentration for RO system design.

The water resources on Earth are classified according to the feed concentration. Brackish water, which is defined based on its range of salinity between seawater and freshwater, includes river water, groundwater, and surface water [18]. The concentration of brackish water was the lowest among the other water resources. Thus, the energy consumption required for the separation of brackish water was the lowest. The low energy consumption makes brackish water desalination technologies attractive because freshwater can be supplied at an affordable price [11, 12, 19, 20]. However, brackish water resources are insufficient to supply all human populations [9, 21]. Thus, it is inevitable to find more abundant water resources for a sustainable supply to human society.

Hypersaline water is defined as saline water containing higher concentrations than seawater, so it has a very high osmotic pressure because of its high concentration. The high osmotic pressure becomes a technical barrier for treating hypersaline water by RO because a very high operating pressure should be applied to the feed solution. This pressure required for hypersaline water treatment often exceeds the pressure limitation of commercial SWRO membranes of less than 81 bar [22, 23]. In addition, the high energy consumption for the treatment of hypersaline makes the hypersaline water desalination infeasible technology because it is not economically favorable compared to other water resources such as brackish water and seawater [24–26]. However, many SWRO plants discharge large amounts of highly concentrated brine, causing significant environmental problems [27]. Therefore, the requirement of brine treatment systems for zero liquid discharge (ZLD) has increased, and hypersaline water treatment by RO has been investigated by many researchers [22, 25, 28]. In addition, since shale gas produced water [29, 30], landfill leachate [31], and flue gas desulfurization wastewater [32] contain very high salinity, effective technologies for the treatment of these types of high-salinity water should be developed. Recently, many new RO processes, such as draw solution assisted RO [33], osmotic-enhanced dewatering [29, 34], and osmotically assisted RO [35] have been developed. These processes can change from a single-stage RO system applying high pressure to a two-stage RO system with low pressure. Therefore, these processes can be effectively used to treat hypersaline water. Nevertheless, the high energy consumption and high cost are major barriers that prevent hypersaline water desalination from being widely utilized as a main water resource for water supply.

Seawater accounts for more than 97% of all water resources on Earth. In water-scarce regions such as the Middle East and North Africa (MENA), seawater desalination is regarded as a permanent solution for water scarcity problems [36, 37]. In these regions, the need for desalination is tremendously increasing [38], and desalinated water is not only used for drinking but also for irrigation and industries [39]. It has been reported that recently installed large-scale SWRO plants can produce freshwater at prices lower than $0.5/m^3 [40], which is significantly lower than the price of thermal-based desalination and much lower than the price of SWRO plants 30 years ago (higher than $2/m^3) [9]. This has led many policymakers and stakeholders to invest in SWRO technology for desalination systems [9, 36, 37]. However, although the energy efficiency of SWRO is the highest among other desalination technologies with high technological maturity [41], it is still necessary to reduce the energy consumption further because the overall desalination capacity installed worldwide is more than 90 million m^3/day [36], and it is expected that the overall energy consumption in all desalination plants will continue to increase consistently. The specific energy consumption (SEC) of the SWRO process has been decreased from 20 kWh/m^3 (in 1970) [42] to only 2.5 kWh/m^3 (in 2010) [21, 43] by developing high-efficiency energy recovery devices (ERDs) [11, 44–46], energy-efficient high-pressure pumps [47–50], and high-performance RO membranes [51–54]. However, the energy consumption of the SWRO process might be reduced further given that the theoretical minimum energy for separation of seawater with a salinity of 35 000 mg/L is approximately 1.07 kWh/m^3 at 50% recovery [21, 45]. In addition, the SWRO process uses a very energy-intensive system [36]. Therefore, it is crucial to minimize the energy consumption of the SWRO system to reduce energy loading in the energy supply system. Recently, global energy crisis has emerged as a major issue, and under these circumstances, low energy consumption in seawater desalination has become very important, and many researchers and engineers have been working to develop innovative ways to minimize the energy consumption of the seawater reverse osmosis (SWRO) process. In addition, the electrical energy required to operate most SWRO plants is supplied by fossil fuels [37]. As power generation by fossil fuels is correlated with greenhouse gas emissions [55, 56], the energy efficiency of SWRO plants needs to be improved for sustainable water supply by desalination systems.

In summary, RO systems for seawater desalination are the most viable technologies for sustainably securing freshwater resources; however, the most recent significant issue in RO systems is energy minimization. RO was developed as an advanced technology for desalination, and research and development to improve the RO system will continue in the future. Recent developments in RO systems have focused on improving energy efficiency for desalination in various feed solutions. In this book, we comprehensively cover the energy consumption issues in the SWRO system from a fundamental understanding to the analysis of the improvement strategy for energy minimization in the future. In Chapter 1, the definitions of osmotic pressure and RO are included to explain the SWRO system. Chapter 2 presents a detailed configuration of existing SWRO plants and the characteristics of each unit process. From the intake to the discharge, all the system units for the SWRO plants are covered in this chapter. In Chapter 3, the methodology for calculating the energy consumption in the SWRO process in a large-scale system is explained. Current energy trends in the SWRO plants, and the manner in which the energy consumption in the SWRO has been reduced are described in Chapter 4. In Chapter 5, the main factors affecting energy consumption are identified from the actual SEC data from the installed SWRO plants, and future strategies for further improvement of energy efficiency are introduced. The actual applicability and future potential of each strategy are discussed in Chapter 6. State-of-the-art technologies for each improvement strategy are also introduced.

1.2 DEFINITION AND OSMOTIC PRESSURE

A semi-permeable membrane allows solvent molecules to freely permeate the membrane, whereas solute molecules do not. When two solutions containing different solute concentrations are separated by a semi-permeable membrane, the net water flux across the semi-permeable membrane is in the direction from the highly concentrated solution to the low-concentration solution by osmosis [57]–the net water flux decreases as the level difference between the two solutions increases. Eventually, there is no net water flux across the membrane at equilibrium, as shown in Figure 1.1. In this situation, the osmotic pressure difference can be measured by the pressure difference arising from the level difference between the two solutions. The osmotic pressure difference becomes larger if the concentration difference between the two solutions is larger. To produce water permeate flux from the highly concentrated solution to the low-concentration solution, a pressure higher than the osmotic pressure difference should be supplied to the solution containing a high concentration. In other words, the osmotic pressure is a hurdle (for water molecules in the highly concentrated solution) to permeate in getting into the low-concentration solution. In an RO system, the osmotic pressure of saline water is very significant because it can be used as a barometer to theoretically estimate the energy required for the separation of water molecules from saline water. Therefore, the osmotic pressure of any saline water should be determined to estimate the theoretical minimum energy required for the separation of freshwater from saline water.

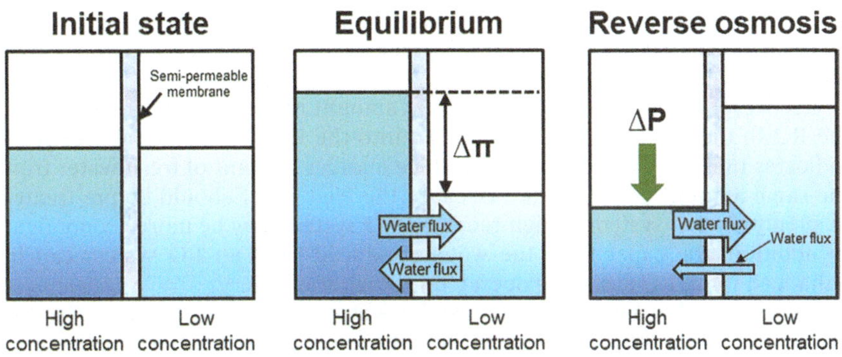

Figure 1.1 Schematic diagram of osmotic pressure and RO.

1.3 PRINCIPLE OF RO

The spontaneous water flux direction in the membrane system is from a highly concentrated solution to a low-concentration solution without any pressure applied to these solutions. As mentioned above, to reverse the spontaneous water flux direction, a hydraulic pressure higher than the osmotic pressure difference between the highly concentrated solution and the low-concentration solution should be supplied to the highly concentrated solution. This principle can be utilized to produce freshwater from saline water with a high-performance semi-permeable membrane, as shown in Figure 1.2, and this is usually called RO.

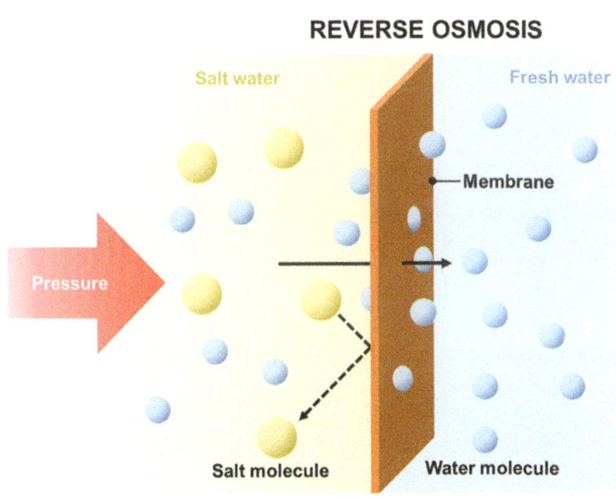

Figure 1.2 Freshwater permeation across the semi-permeable membrane in RO.

Many key terms utilized in RO are summarized in this chapter; it is recommended that readers familiarize themselves with these terms as they relate to current RO systems.

Recovery: Recovery is the ratio of the amount of water permeated through the RO to the amount of total water fed into the RO system. A high recovery indicates that the RO system can produce a larger amount of fresh water from the same amount of feed water. Because the feed water should be pre-treated to supply the RO system, a high-recovery RO system may be more economical. In addition, the capacity of the water produced from an RO system can be enhanced if the RO system is operated for higher recovery.

Permeate: The permeate denotes the water transferred through the RO membrane. The permeate is usually fresh water in an RO desalination system.

Rejection: Rejection is the ratio of the solution concentration of the permeate stream to the concentration of the feed stream. This is one of the main performance indicators of RO membranes. If the rejection is very high, the RO system can produce a very high-quality permeate with a very low salt concentration.

Retentate (concentrate): The retentate is water rejected from the RO membrane. The retentate consisted of salt and water retained in the RO system. Because the permeability of the semi-permeable membrane is much lower than that of water, the retentate has a higher concentration than the feed solution. The concentration and amount of the retentate stream are determined by the rejection of the RO membrane and the recovery of the RO system.

doi: 10.2166/9781789061215_0007

Chapter 2
SWRO desalination plants

2.1 OVERALL PROCESS

Recent seawater reverse osmosis (SWRO) desalination plants are standardized high-efficiency plants. Typical processes of an SWRO desalination plant are illustrated in Figure 2.1. However, specific designs can vary depending on the feed quality, required product quality, and recovery. An SWRO desalination plant is composed of three parts: pre-treatment, RO process, and post-treatment.

Pre-treatment units were employed to remove large particles and solids before the SWRO process. In seawater, various ions/particles/solids/organic matter are dissolved or suspended. Undissolved solids frequently cause fouling (except scaling) on the surface of RO membranes or damage RO membranes. Thus, it is crucial to remove the foulants before the RO process to maintain and secure the RO performance. The pre-treatment process removes undissolved solids in the feed and supplies foulant-free feed to the RO system. The process is composed of intake, dissolved air flotation (DAF) (optional), dual-media filtration (DMF) or MF/UF (microfiltration/ultrafiltration), and cartridge filter (CF).

The RO system can remove dissolved ions from the pre-treated feed and produce freshwater. Chemical species and ions are dissolved in seawater (i.e., total dissolved solids (TDS)), and the concentration of TDS generally ranges between 30 000 and 40 000 mg/L [58]. However, the TDS range can be 45 000 to 48 000 mg/L in the Arabian Gulf. TDS should be rejected through the membrane to obtain fresh water from seawater, and this can be achieved using semi-permeable RO membranes. However, the feed has an osmotic pressure of 20–30 bar due to high TDS, and the osmotic pressure should be overcome to separate water and salts across the SWRO membranes. Thus, the pre-treated feed is pressurized by a high-pressure pump (HPP) and fed to the SWRO membranes.

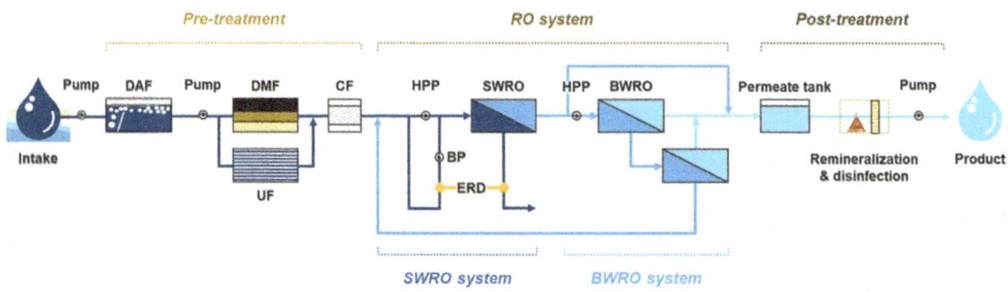

Figure 2.1 Scheme of SWRO desalination process. A feed is pre-treated by several processes, including DAF (optional), DMF/UF, and CF. The pre-treated feed is sent to an RO system composed of an SWRO system and a BWRO system (optional). The permeate is processed by remineralization and disinfection, and the product is distributed.

When the hydraulic pressure of the feed is higher than the osmotic pressure, water is produced on the permeate side, and the rejected feed is sent to the outlet. Multiple SWRO membrane modules (e.g., six–eight modules) are installed in a pressure vessel (PV) and recover 40–50% of feed as permeate. The rejected feed is called concentrate/brine, and an energy recovery device (ERD) recovers hydraulic energy and delivers it to the part of the feed associated with a booster pump (BP). The SWRO permeate can be used directly, but further desalination can also be performed using brackish water reverse osmosis (BWRO) membranes.

Post-treatment is necessary to stabilize and disinfect water. As divalent ions (e.g., Ca^{2+} and Mg^{2+}) are highly rejected through SWRO membranes, desalinated water has low hardness. In addition, both the pH and alkalinity are low. Thus, desalinated water is remineralized with useful minerals to secure the water distribution system from corrosion and maintain water quality. Furthermore, disinfection using chlorine of ultraviolet (UV) irradiation is required to prevent biological activity and contamination.

Because each compartment is associated with one another, the application of new desalination technologies to the current system would be a challenge. In addition, engineers in the industry prefer to use conventional technologies rather than unproven technologies. However, the design of SWRO desalination plants has advanced further with decades of experience.

2.2 SWRO INTAKE

The intake system of the SWRO plant is a critical and significant aspect of the process. The intake unit may occupy more than 20% of the capital cost while still occupying a significant portion of the operating expenses [59]. A good intake system would not only meet the water quality required for the SWRO plant but would have less impact on the ecosystem. The types of SWRO intake include open water intake and subsurface intake. The selection of intake type is based on a thorough assessment of sea conditions and environmental concerns.

2.2.1 Open water intakes

Open water intake is the most widely used intake type because of its high intake volumes, suitable for large SWRO plants, providing an opportunity for co-location with power plants. They can be easily installed at any location without serious consideration. Open intake is classified as surface open intake or submerged open intake.

Surface open intake incorporates active screen systems to eliminate debris, aquatic life, and trash from reaching the SWRO line. Traveling water screens are furnished with both revolving and wire mesh panels, with panel mesh sizes varying from 9.5 to 13 mm. The spinning of wire mesh panels allows high-pressure water to remove accumulated debris from the screens. Smaller mesh sizes can be used seasonally or regionally to remove eggs and larvae, but this could cause operational difficulties as debris and marine life increase significantly.

The feed is collected by a submerged open intake 2–6 m above the seabed. Passive displays are mounted in the submerged open intake (fine: 0.5–10 mm, coarse: 50–300 mm). There is no movement or velocity limit, which is a behavioral barrier that keeps aquatic life and sediment out of intake pipes. The intake head can be fitted with a coarse velocity cap. The carriage pipe is then used to store water offshore. Open intake can easily be scaled up or down to meet capacity requirements. However, impingement/entrainment and heavy organic load make it vulnerable. Because the RO membrane is sensitive to various types of foulants, SWRO plants with open water intake systems normally have more complex pretreatment processes. As a result, pretreatment units that can manage any water quality must be built to remove silts, organics, and microbes that cause clogging and fouling of plant equipment and membranes. Pretreatment failure may affect the quality of the permeate.

2.2.2 Subsurface intakes

Subsurface intake collects seawater through sea wells (e.g., vertical, horizontal, slant, or radial), infiltration galleries, or other locations below the seabed. Subsurface intake is more suitable when geological conditions are below the seabed owing to their natural filtration ability. Natural filtration in subsurface intake is similar to sand filtration during physical and biological removal. Subsurface intake can remove microorganisms such as bacteria, algae, and biopolymers (proteins and polysaccharides) from the intake water. Thus, the intake system can reduce clogging and fouling of SWRO pretreatment units/membranes. The feed is relatively clean and thus requires less chemical treatment, such as antiscalants and coagulants. Furthermore, subsurface intakes pose less risk to environmental and marine life because the productivity of the ocean is believed to decrease with increasing depth [60]. Therefore, environmental problems such as impingement and entrainment encountered in open water intakes are mitigated by subsurface intakes.

However, subsurface intakes are not without challenges. For example, subsurface intakes require extremely favorable hydrogeological conditions, which are usually challenging to find in the vicinity of SWRO plants. It is commonly used for small SWRO plants. Debris is retained on the ocean floor in the well area, and beach erosion could hinder the performance of the subsurface pretreatment due to the corrosion of the filtration layer over time. Subsurface waters are lower in dissolved oxygen (DO) (between 0 and 1.5 mg/L) and high in hydrogen sulfide, which are not rejected by the RO membrane [61, 62]; hence, additional cost would be required to reaerate produced water. The release of low DO brine would cause oxygen depletion, causing harm to marine life around the brine discharged area, and significant maintenance efforts are required for periodic dredging or replacement of the upper portion of the intake filtration media.

2.3 PRE-TREATMENT

2.3.1 Dissolved air flotation

In the SWRO process, dissolved air flotation (DAF) is used as a clarification technique to remove solid particles before passing the effluent to either DMF or UF units. The principle of DAF involves the attachment of air bubbles to particles or flocs and then moves them to the surface of the water. The DAF system is usually divided into two compartments, the contact and separation compartments (Figure 2.2). It is paramount to allow collision of solid particles and bubbles in the contact compartment. The attachment of air bubbles to solid particles is called floc-bubble aggregation. Floc-bubble aggregates, and the remaining free bubbles are transferred to the separating compartment where free bubbles and floc-bubble aggregates float to the surface of the water. As the concentration of floc-bubble aggregates increases on the surface of the water, sludge is formed, collected, and removed. Clarified water at the bottom of the DAF unit (sometimes called a subnatant) was collected and transferred to DMF or UF for further pretreatment.

Bubbles (10–100 μm) are generated by saturating pressurized recycle flow with air. The rapid release of pressure via nozzles or special valves at the

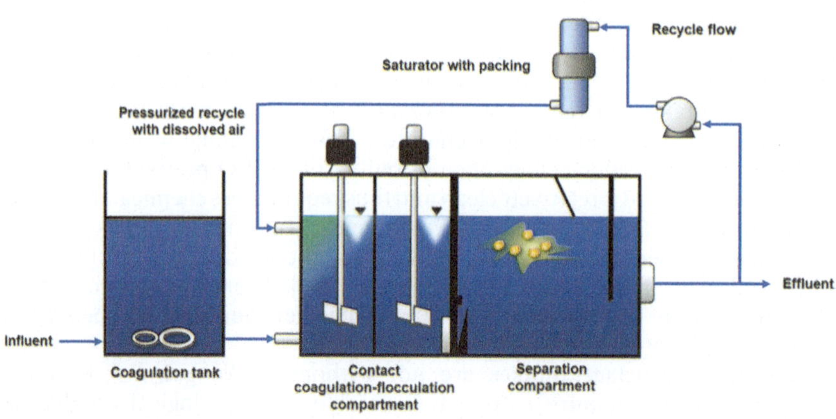

Figure 2.2 Dissolved air flotation unit.

bottom of the influent entrance spontaneously produces bubble nuclei due to the difference in pressure between the nozzles/valves and supersaturated water. The steady-state size of a bubble depends on the saturator pressure or air injection flow rate. High injection pressure results in small bubbles; however, a pressure point is reached where the increase in injection pressure does not correspond to a reduction in bubble size [63]. Bubbles are formed in the DAF unit via nucleation (homogeneous and heterogeneous) and growth. Homogeneous nucleation is a process of bubble formation within water when the difference between the surrounding pressure and dissolved air pressure is more than 100 bar [64]. The critical bubble diameter (d_{cb}) in homogeneous nucleation is projected using Equation (2.1):

$$d_{cd} = \frac{4\sigma}{\Delta P} \tag{2.1}$$

where σ represents the surface tension of water and ΔP represents the pressure difference across the nozzle or injection valve. However, homogeneous nucleation is not expected in a DAF unit. On the contrary, heterogeneous nucleation is the process of bubble formation within pre-formed air gas pockets in the vicinity of surface cracks and imperfect solid particles. Heterogeneous nucleation occurs during the supersaturation of water [65]. Consequently, the bubbles grow by (1) uptake of air from water, (2) hydrostatic pressure decrease due to the upward movement of bubbles, and (3) merging or coalescence of bubbles.

Floc–bubble aggregates are formed by the entrapment of bubbles in a large floc, the growth of bubble nuclei on or with the floc, and collision and merging with the floc. The electrical charge interactions and attractions with van der Waals forces must be reduced for particles to attach to bubbles effectively. This is because most of the suspended particles are negatively charged, and bubbles are also thought to be negatively charged because of the ability of anions in water to attach more easily to the air bubble. For the particles to float, the charge of the particles must be neutralized, and the hydrophobicity must be increased [66]. Prior physical/chemical pretreatment of DAF influent by the addition of cationic surfactants or polyelectrolytes enhances floc–bubble aggregation [67–69].

The fundamental mechanism of solid particle removal in DAF is flotation. Unlike the sedimentation tank, DAF is more effective at removing low-density particles that cannot be removed from the sedimentation tank. Thus, DAF is more appropriate as a pretreatment for the SWRO process for algal-laden seawater rich in organic matter. Moreover, DAF does not necessarily require the formation of large flocs to achieve optimal performance. Therefore, a lower coagulation dose and shorter flocculation time compared to the sedimentation process are required. Nevertheless, coagulation is an important pretreatment step in DAF. Coagulation helps to transform the negative charge of particles to enable bubble attachment. Examples of seawater coagulants used for pretreatment are ferric chloride, aluminum sulfate, and polyaluminum chloride. Similarly, design parameters such as the hydraulic loading rate, solids loading rate, and air-to-solids ratio are essential for the optimum performance of the DAF system. This predicts the quantity of water and air that should be supplied to the DAF system simultaneously for efficient treatment and operation. The hydraulic loading rate is expressed in terms of the relationship between the DAF surface area and the total flow rate (influent and recycling). On the contrary, the solid loading rate is expressed with respect to the relationship between the DAF effective surface area and the total amount of solids entering the DAF unit. DAF can be operated efficiently at higher surface loading rates (SLRs). As in the sedimentation overflow rate, hydraulic loadings were used to represent the rate and size of the DAF unit. A conventional DAF unit is designed with hydraulic loading between 5 and 25 m/h, while newly designed units have hydraulic loadings between 15 and 30 m/h or more.

Owing to the flotation mechanism in DAF, it is very effective for removing low-density solid particles. In the SWRO process, DAF is applied because of its suitability for removal in algae-laden water [70, 71]. The occurrence of harmful algal blooms (HABs) in the Gulf of Oman and the Arabian Gulf in 2008–2009 resulted in the closure of numerous SWRO plants. This intensified the necessity for DAF as a pretreatment unit in the SWRO process for the removal of algae and algal-related organic matter. The use of DAF technology has increased dramatically since 2009, with almost all newly installed plants equipped with DAF units. In a report by Veolia, the use of DAF in the Fujairah II SWRO plant enabled the successful operation of the plant, even under severe HAB occurrences [72]. Gallou *et al.* reported 99% removal of algae by coagulation-AquaDAF™ followed by DMF in Al-Dur, Bahrain [73].

Appropriate pretreatments prior to DAF treatment would increase its effectiveness in handling algae, turbidity, organic matter, and color. The optimal coagulation dose and pH adjustment can affect the particle charge, and thus, the bubble can quickly attach to them. It is important that the coagulants used for pretreatment do not increase the hydrophilicity of the particles. Flocculation in DAF is quite different from flocculation during sedimentation. In sedimentation pretreatment in the SWRO process, the goal is to have a large floc that can easily settle to the bottom of the sedimentation tank. On the contrary, the floc in DAF is expected to be smaller for quick flotation. The preferable floc sizes were within 25–50 μm. Overall, DAF had superior solid particles and turbidity removal over sedimentation.

The DAF requires high energy and maintenance costs. Saturated pressurized recycle flow and aeration significantly contribute to the power cost of the unit. The average energy consumption was reported to be 0.05–0.075 kWh/m^3 of water treated [74]. Another recent work presented estimated SEC under varying recycling rate and aeration saturation pressure conditions, with a value ranging from 0.01 to 0.06 kWh/m^3 [75]. Although certain levels of algal cell concentration and turbidity are present in seawater year-round, occurrences that can be classified as HABs are seasonal events during the spring and fall seasons. Extreme cases such as those of the Gulf of Oman and the Arabian Gulf in 2008–2009 are infrequent. This affects the overall efficiency of DAF as higher efficiency is obtained in heavy-laden algae water compared to low-algae-level waters. Hence, research attention is needed to increase the efficiency of DAF regardless of algae concentration and to reduce the energy required for bubble formation.

2.3.2 Dual media filtration

A wide area of the filtration bed is used in the depth filtration process. The efficiency of the depth filtration is determined by the size and density of the medium. Backwashing is an efficient way to improve the performance of depth filtration owing to the density of the medium. Therefore, depth filtration is a better option for seawater desalination than surface filtration pretreatment. The principle is to implement DMF with different densities to boost the efficiency of depth filtration (e.g., filtration speeds, operational runs, and minimal backwash water usage). This concept is called DMF (Figure 2.3).

For high suspended solids and turbidity reduction, various SWRO plants use DMF as a primary pretreatment for seawater/brackish water feed. DMF is a mechanical process that removes turbidity, organic matter, color, and odor by straining, sedimentation, impaction, interception, adhesion, adsorption, flocculation, and sometimes biological growth. Sand, calcites, anthracite, and activated carbon are some of the media used in DMF units. Calcite medium is widely used because it is abundant in nature, inexpensive, and helps to improve pH levels by neutralizing water. Because it is lighter than sand and has a high carbon content, anthracite is commonly used in DMF systems to remove suspended solids and turbidity while resisting chemical attack. Activated carbon media can eliminate large amounts of organic matter, ammonia, and

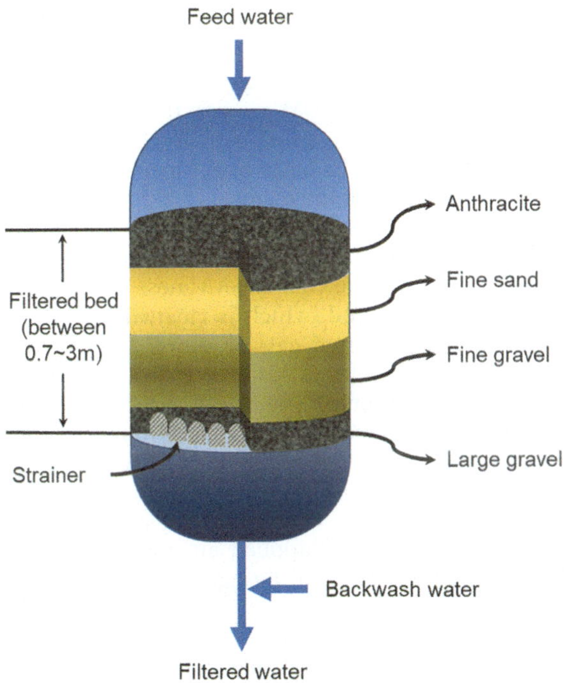

Figure 2.3 Layout of DMF.

disinfectants. Because of the high cost of anthracite, it is usually mixed with a sand bed. Anthracite ranging in size from 1.5 to 2 mm, with an average size of 0.5 Â mm, is placed on top of the sand bed.

DMF incorporates both depth and surface filtration techniques. DMF is more efficient than surface filtration alone. This is because the entire depth is used, provides better quality performance, and has a high flow rate. DMF is more cost-effective because it can be used with chemical coagulation, lowering the overall process cost of the system. Additionally, in the presence of high organic matter concentrations or turbidity spikes, coagulants are needed to reduce the footprint and ensure adequate permeate consistency (SDI 5). In the DMF unit, either inline or full-scale coagulation was carried out. Coagulants, such as ferric chloride, are widely used in SWRO plants.

Media properties, such as size distribution, shape, density, and porosity, are crucial in determining the filtration performance of DMF. The amount of head loss that accumulates during a filtration run is affected by the size of the media. The smaller the media, the smaller the pore opening, allowing water to permeate and aids foulant removal. The efficiency of the DMF unit improves if the pore opening is minimal. Small pore openings increase head loss through the media. The size of the medium was specified as the effective size. The effective size is defined as the size of the openings that allow 10% of the medium by weight to pass. The uniformity of the media is represented by the uniformity coefficient. The uniformity coefficient is measured as the size opening that allows 60% of the medium to pass through, divided by the size opening that allows 10% of the medium to pass through. The uniformity coefficient describes the spectrum of media sizes, whereas the effective size describes the average media size. The ideal uniformity coefficient of DMF is less than 1.7. Media shape is also important because it affects the fixed-bed porosity, filtrate head loss, filtration efficiency, and ease of filtration. The DMF backwash efficiency is dependent on the shape, density, and porosity of the media.

Backwashing of DMF removes the suspended materials that accumulate on the filter media during filtration. The ease with which DMF can be washed by backwashing contributes to the cost-effectiveness of the process. However, unlike the single filtration method, which is cleaned easily by backwashing, cleaning the DMF by backwashing can be more difficult at times. This is because the top layer of DMF is normally fine and can be identical in size to the attached materials. The fine topmost layers could be washed with suspended materials without caution. Increased head loss through the media up to the available lower-level limit, degradation of filtered water quality, and exceeding the maximum time limit are all indicators of backwashing in DMF. DMF is usually backwashed with water or water aided by air scouring after 24–36 h in SWRO pretreatment measures. Supplied air may be applied before water backwashing, or both air and water may be applied concurrently in air-scouring supported backwashing.

Although the DMF method uses fewer resources, it has a low level of organic and biofouling mitigation. If the organic load and turbidity of the SWRO intake water are high and not removed prior to the DMF unit, the DMF unit will become clogged, resulting in short filter runs and a filtrate with a high risk of fouling the SWRO membrane. Unfortunately, in an environment with a high organic and microbial load, DMF pretreatment is insufficient to produce water of adequate quality to minimize fouling on SWRO membranes [76]. The filtration rates should be shortened to ensure the adequate use of DMF with high organic and microbial loads. A shorter rate would result in less clogging of the media (this is still subject to the physical properties of feedwater) [77]. Nevertheless, the shortening of the filtration rate requires a corresponding increase in the active surface area of the media. This could lead to an increase in the footprint and construction costs of the SWRO pretreatment system. Plant operators have attempted to improve the performance of DMF by improving the coagulation strategies.

Several studies have examined the efficiency of DMF under various conditions in the literature [78–82]. Overall, the use of coagulation prior to DMF showed improved performance in terms of turbidity and organic removal compared to DMF alone, validating the need for some pretreatment steps for DMF unit stability.

2.3.3 Membrane-based pretreatment

The SWRO process has seen improvements in all aspects of the process, including the RO membrane, module design, process optimization, and pretreatment. Pretreatment is an essential component of the SWRO process. This is because, regardless of how far other areas of the SWRO process have progressed, their performance is still heavily reliant on feed quality due to potential issues, including membrane fouling caused by poor feed quality. Fouling in the SWRO process can be avoided by using an effective pretreatment process. Furthermore, with a given hydraulic pressure, a high-quality feed will sustain the permeate flux. Therefore, the need for implementing appropriate pretreatment processes is a viable way to increase the energy efficiency of SWRO desalination. Traditional pretreatment technologies, such as coagulation/flocculation, disinfection, DAF, and DMF, have been previously used in SWRO plants to pretreat seawater feed. However, these methods have not been able to overcome the difficulties of membrane fouling. This has led to the recent implementation of advanced pretreatment technologies, such as MF, UF, and nanofiltration (NF).

2.3.3.1 Microfiltration

MF pretreatment using membranes with pore sizes in the range 0.1–0.35 μm [83] was introduced as a pretreatment unit in the SWRO process as a submerged MF process usually installed in the SWRO process owing to its fouling resistance and ability to withstand different seawater feed conditions. The MF process can effectively remove particulate matter and microbes from the feedwater, thus preventing fouling on the RO membrane. Unfortunately, during MF pretreatment, a large amount of foam is formed because of the protein content of marine microbes [84] and the MF membrane is easily fouled. Therefore, for the efficient and effective operation of the MF system, an adequate pretreatment step such as coagulation/chlorination is necessary for better removal of organic and inorganic materials to mitigate fouling on the MF membrane [85, 86]. Significant efforts have been made to improve MF membranes to enhance flux, rejection, and reduce fouling and cost. These efforts are centered on improving MF membrane hydrophilicity, incorporating into the membrane matrix, or using inorganic materials [87].

2.3.3.2 Ultrafiltration

The UF process has gained more applicability in the SWRO process than the MF process. This is due to its higher operational flexibility and considerable balance between water transport and foulant rejection in difficult water conditions. The UF membrane pore size is smaller than that of the MF membrane, with pore sizes ranging from 0.01 to 0.05 μm [83]. UF can eliminate a wide range of pollutants present in seawater feed, such as algae, bacteria, fungi, silt, and some organic matter. The fouling potential of the RO membrane by permeate from the UF system is lower than that of the MF system. However, as in the MF system, the efficiency of the UF system is improved by coupling the UF system with conventional pretreatments such as coagulation, DMF, or DAF. A consistent UF permeate of <3 SDI can be achieved by coupling UF with the appropriate coagulant during HAB [88]. A lower SDI of 0.5 was reported when the UF system was coupled with coagulation [89]. Inappropriate pretreatment is detrimental to the UF membrane because foulants can accumulate on the membrane surface and within its pores. The impact of foulants on the UF membrane can be minimized by developing new UF membrane materials or modifying the UF membrane surface.

2.3.3.3 Nanofiltration

Even though significant improvements in feedwater quality have been achieved using MF and UF systems, RO membrane fouling is still a significant problem in the SWRO process, especially with difficult feedwater. In addition, MF and UF have poor ionic species rejection. Hence, the permeate of these processes has numerous low-molecular-weight organics and scaling fouling potentials. Thus, the NF unit was introduced. An appropriate NF membrane has a molecular weight cut-off (MWCO) below 1000 Da and can effectively reject both charged and uncharged ions [90]. Today, the research trend is to use NF pretreatment instead of MF and UF. Although NF requires higher pressure due to the tightness of the NF membrane pores, NF, however, produces a lower saline permeate. Lower feed salinity passed on to the RO membrane would, in turn, reduce the pressure required to transport water across the RO membrane. Therefore, utilizing NF for SWRO pretreatment reduces the RO pressure requirement due to low salinity and organic foulants and increases RO recovery and water flux [91, 92]. However, effective and consistent operation of the NF process can only be achieved by adequate pretreatment of raw feedwater; similar to the RO unit, the NF system is susceptible to fouling. Therefore, to ensure proper pretreatment, the NF system is coupled with other membrane processes, such as MF or conventional pretreatment processes, to improve the quality of NF feedwater conditions. Similarly, recent developments in membrane materials (inorganic, ceramics, etc.) have been shown to be alternative membrane materials for the NF process. This is because these new membrane materials can tolerate harsh feedwater conditions, and have self-cleaning properties (stimuli-responsive materials).

2.3.3.4 Inline coagulation using advanced coagulation process

Owing to the propensity of membrane-based pretreatment for fouling, adequate pretreatment is required. A common practice is to combine a membrane-based pretreatment unit with a conventional coagulation process. However, conventional coagulation is often followed by flocculation and sedimentation tanks to remove flocs and sludge before the membrane unit. This increases the footprint of the SWRO process. Therefore, the introduction of inline coagulation eliminates the need for a sedimentation process, thus reducing the SWRO footprint. Recently, an advanced coagulation process has been introduced in the SWRO process. Inline advanced coagulation technology using liquid ferrate at low concentrations is an excellent technology because of the combined oxidation, disinfection, and coagulation properties. This inline technology saves both capital and operating costs that would otherwise be spent on energy and chemicals. The addition of oxidation aids in removing organic components from the feedwater without the need for physical separation. As organic and biological components can be mineralized, the effluent from this phase would have limited environmental effects. Superior microbial and organic removal was documented when an advanced coagulation pretreatment process was combined with a membrane process [93, 94].

2.3.4 Cartridge filter

Cartridge filters (CFs) are pressure filtration devices used for the removal of suspended solids and chemicals for which total concentrations are less than 100 mg/L. Filtration by CFs is mostly outside-in; therefore, filters need to be mechanically strong to accommodate changes in pressure during filtration operation. Hence, the CFs are enclosed within a housing/casing to withstand such pressure changes. In the SWRO process, CFs are often used as a pretreatment step before the RO system. These filters are available in various lengths and diameters, and materials (e.g., pleated membranes, woven, and non-woven).

CFs are categorized into two types: surface and depth cartridge filters. Surface CFs remove contaminants, mainly by size exclusion. Here, contaminants larger than the pore size of CF are restricted from entering the filter media. Depth CFs, on the other hand, remove contaminants by Brownian transport. Here, the multiple layers and thickness of the filter provide a tortuous path for the contaminants, and eventually, they are trapped within the filter. With depth CF, contaminants are often smaller than the pore size of the filter. Regardless of the CF type, contaminants are either collected on the filter surface or within the pores until the filter clogs. Filters are sometimes designed to possess cleaning cycles; however, in the case of heavy fouling, the filters can be chemically cleaned (if recommended by the manufacturer) or replaced.

2.4 RO SYSTEM EQUIPMENT

The feed after the pre-treatment unit is fed into the RO system to produce freshwater. A schematic of the RO system is shown in Figure 2.4. Because water permeation in the RO system is conducted by pressurization, the main body of the RO system consists of a HPP for pressurization and a PV for membrane housing. It is desirable to recover the remaining pressure energy in the concentrate stream to maximize the energy efficiency of the RO system. Thus, an ERD should be installed between the outlet stream of the concentrate and the inlet stream of the feed to transfer the remaining pressure energy from the concentrate to the feed; then, the concentrate stream is discharged from the RO system after the ERD. Owing to the imperfect recovery of the remaining pressure and the pressure drop during the RO system, the feed stream after the ERD is insufficient to operate the RO system continuously. The insufficient pressure on the feed stream is compensated for by a high-pressure pump, usually called a BP. This is the basic structure of an RO system. Hence, the energy consumption of the RO system is determined by a HPP and BP. To run the RO system effectively, it is imperative to use a HPP and an ERD with high energy efficiency.

2.4.1 High-pressure pump

A positive displacement pump can achieve an energy efficiency of over 90%; hence it can be utilized as a high-pressure pump in seawater desalination systems requiring high energy efficiency. In some cases, the positive displacement pump can even show a very high pump efficiency of approximately 97% [36]. Many small desalination plants have taken advantage of the high efficiency of positive displacement pumps [36]. Although a positive displacement pump has some disadvantages, such as complex structure and difficulty in operation, its high energy efficiency provides substantial merit for minimizing the energy consumption of desalination plants. Thus, the use of positive displacement pumps was recommended for many new desalination plants in the Canary Islands [95, 96], as well as technologies such as batch RO [97, 98] and CCD [99]. However, a positive displacement pump cannot be easily used in large-scale

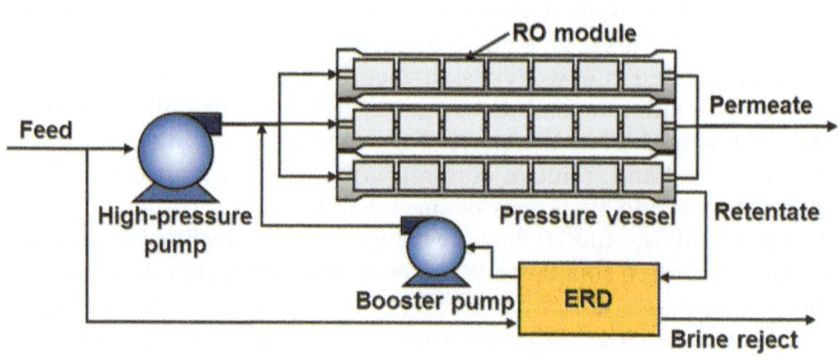

Figure 2.4 Schematic of RO system.

desalination systems of over 1000 m^3/day. The main reason for this is that a large positive displacement pump is costly, increasing the water production cost immensely.

On the contrary, centrifugal pump is widely utilized in many large-scale desalination plants such as the Sydney Water, Perth I and II, Adelaide SWRO desalination plants in Australia, and many desalination plants in Spain and the Middle East [36]. Centrifugal pumps can be easily and cheaply manufactured for a large volume of feed supply compared to a positive displacement pump, which is the main reason for its wide utilization in large-scale desalination plants. Although centrifugal pumps have lower pump efficiency than positive displacement pumps [100], the advantage of using a centrifugal pump is quite beneficial for the economics of operating desalination plants. In addition, the energy efficiency of a centrifugal pump can be improved by expanding its size because a multistage centrifugal pump design in the large-size desalination plants could be operated more effectively than the single-stage pump. The practical limitation of a centrifugal pump is its efficiency of approximately 90% [36]. In many recently constructed large-size desalination plants with over 100 000 m^3/day capacities, high-efficiency, large-size centrifugal pumps are used, and the efficiency of the pumps is higher than 85% [36].

2.4.2 Energy-recovery device

Many of the SWRO plants are operated under 40–50% recovery because of the limitation of the maximum sustainable pressure on the SWRO membrane (<83 bar) [101]. Thus, a high amount of concentrate is discharged at high hydraulic pressure. Such hydraulic pressure in the concentrate is recovered by the ERD application and can reduce the energy consumption of the RO system. The development of ERDs has been one of the main contributors to energy reduction in SWRO plants in recent decades [11].

Different types of ERD can be applied depending on the design of the SWRO plants. ERDs are classified into two groups: turbine (or centrifugal) and isobaric ERD. Turbine ERD includes a Francis turbine (FT), Pelton turbine (PT), turbocharger (TC), and isobaric ERD, which includes dual work exchanger energy recovery (DWEER) and pressure exchanger (PX). However, most SWRO plants employ ERD in PT, DWEER, or PX.

PT recovers the hydraulic pressure of the concentrate by rotating the shaft connected between the wheel and the HPP (Figure 2.5a). PT is relatively easy to operate at a relatively low cost, and its size is more compact than isobaric ERDs [102]. However, as the hydraulic energy of the concentrate is converted into mechanical energy and again to hydraulic energy, the energy efficiency of PT is lower than that of isobaric ERDs. As a result, the use of PT was the mainstream ERD application in the early 2000s, but its application dramatically decreased with the development of isobaric ERDs.

DWEER directly delivers the hydraulic pressure of the concentrate to the feed by operating the pistons (Figure 2.5b). At the two sides of the device, one is filled with the feed with low pressure, and the other is filled with the concentrate with high pressure. The hydraulic pressure is exchanged when

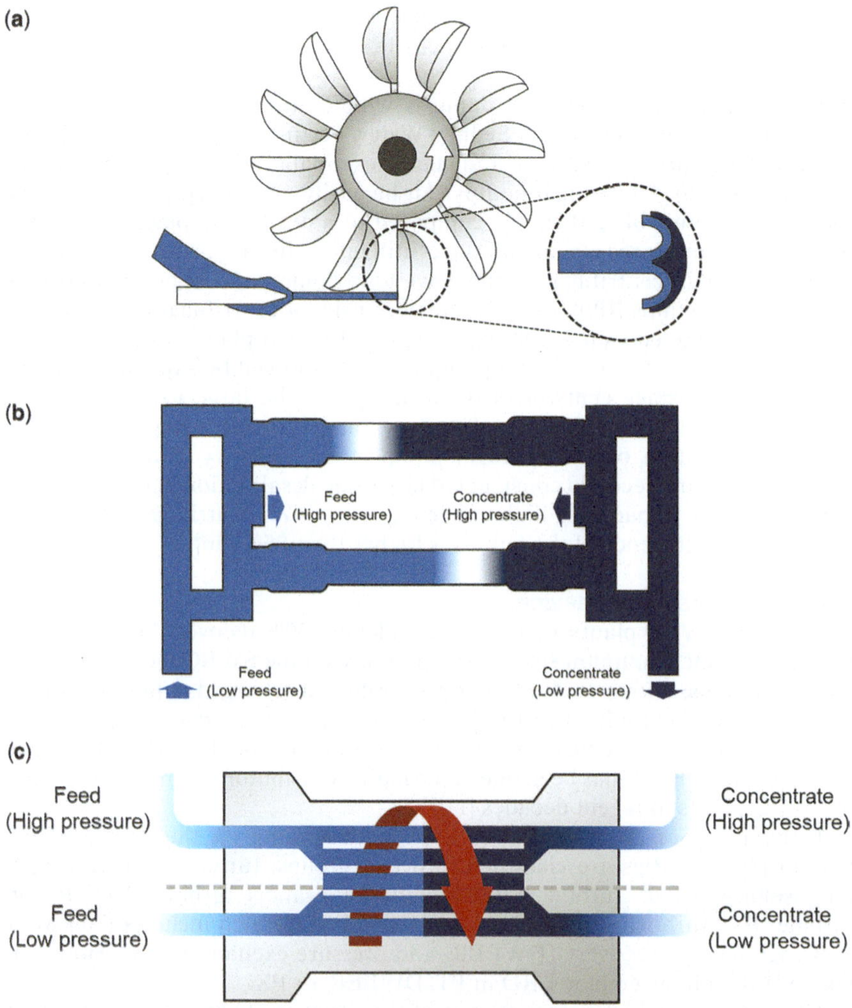

Figure 2.5 Types of ERD: (a) PT, (b) DWEER, and (c) PX.

the pistons move back and forth. The direct exchange of hydraulic pressure improves the energy efficiency of the DWEER over turbine ERDs. However, moving parts in DWEER would require intensive maintenance, which is not favorable for operators.

The PX is equipped with a rotor to deliver the hydraulic pressure of the concentrate to the feed directly (Figure 2.5c). The rotor rotates when it is filled with low-pressure feed, and the low-pressure feed in the rotor is exposed to the high-pressure concentrate. The feed is then pressurized by the concentrate and sent to the feed line of the RO system. The unpressurized concentrate is pushed by low-pressure feed and discharged. It has been reported that PX exhibits superior energy efficiency compared to other commercial ERDs, leading to low-energy SWRO desalination. Despite having the highest energy efficiency, high-salinity concentrates can be mixed with low-salinity feed. The increase in feed salinity slightly increases the SEC and lowers the permeate quality.

Different types of ERD have been employed in SWRO desalination plants. In particular, high-energy-efficient ERDs are adopted, as they are fundamental in reducing the SEC of RO systems. Currently, as PX is highly energy-efficient, low-cost, and stable, it has been widely applied to SWRO desalination plants. However, PX might be substituted by a new type of ERD if several improvements are launched moving forward.

2.5 RO MEMBRANES

RO membranes are the main equipment used in SWRO desalination. This allows water molecules to permeate but rejects salts toward the permeate side. As a result, the performance of RO membranes determines the performance of the RO system (e.g., water quantity and quality). Different RO configurations can be adopted based on the characteristics of the membranes. Thus, the selection of RO membranes is critical, depending on the target design.

The performance of RO membranes is indicated using water permeability (A) and salt permeability (B). Because membrane manufacturers provide the performance of membranes under test conditions, water and salt permeability should be calculated using several mathematical equations by utilizing Equations (2.2) and (2.3):

$$J_w = A\left[\left(P_f - P_p\right) - \left(CPF \times \pi_f - \pi_p\right)\right] \tag{2.2}$$

$$J_s = B\left(CPF \times C_f - C_p\right) \tag{2.3}$$

where J_w and J_s are the water and salt fluxes, P_f and P_p are the hydraulic pressures for the feed and permeate, respectively; π_f and π_p are the osmotic pressures for the feed and permeate, respectively; C_f and C_p are the concentration for the feed and permeate, respectively; and CPF is the concentration polarization factor. Thus, several SWRO and BWRO membranes were calculated and are listed in Table 2.1. The water permeability values for the SWRO membranes were approximately 1–2 L/m^2 h bar, and those for BWRO membranes are 3–6 L/m^2 h bar (Figure 2.6).

RO membranes can be classified as depending on water and salt permeability. High-rejection membranes are those with low salt permeability values. It rejects salts with high efficiency, but the water permeability is relatively low. In contrast, high-flux membranes exhibit high water permeability, but their salt permeability is low. Although high water permeability and low salt permeability are preferred to improve the efficiency of desalination, it is difficult to achieve both objectives.

Table 2.1 *A* and *B* values for commercial SWRO and BWRO membranes.

Type of RO membrane	Manufacturer	Model	Water permeability, A (L/m² h bar)	B (L/m² h)
SWRO	LG Chem	SW400GR	1.25	4.36×10^{-2}
		SW400R	1.52	5.20×10^{-2}
		SW400ES	2.36	1.04×10^{-1}
	DuPont Water Solutions	SW30XHR–400i	0.99	4.22×10^{-2}
		SW30HRLE–400i	1.25	5.82×10^{-2}
		SW30XLE–400i	1.52	6.94×10^{-2}
	Toray	TM820K–400	0.98	3.11×10^{-2}
		TM820C–400	1.10	6.22×10^{-2}
		TM820M–400	1.19	5.34×10^{-2}
		TM820E–400	1.28	7.11×10^{-2}
		TM820V–400	1.56	6.79×10^{-2}
	CSM	RE8040-SHN 400	1.08	6.34×10^{-2}
		RE8040-SHA 400	1.25	7.28×10^{-2}
		RE8040-SHF 400	1.63	1.12×10^{-1}
BWRO	LG Chem	BW400R	3.30	1.48×10^{-1}
		BW400ES	5.36	1.48×10^{-1}
	DuPont Water Solutions	BW30–400	3.29	1.92×10^{-1}
		LE–400	5.86	1.25×10^{-1}
		ECO–400i	5.89	1.21×10^{-1}

Basic performances for SWRO membranes were obtained in the condition of 32 000 ppm NaCl and 5 ppm boron at 25°C (77°F), 800 psi (55 bar), pH 8, and 8% recovery. Those for BWRO membranes were evaluated with 2000 ppm NaCl at 25°C (77°F), 225 psi (15.5 bar), pH 7, and 15% recovery.

Several RO modules are equipped with thicker spacers to reduce the fouling propensity. Fouling can be easily formed when the feed channel (i.e., space between membrane leaves) is narrow. Generally, 28 mil (0.71 mm) spacers are employed in membrane modules, but 34 mil (0.86 mm) spacers can be used to mitigate fouling formation. RO modules with 34 mil spacers can be employed in the region where the feed sources contain high foulants, such as organic matter.

Because the key of SWRO technologies is RO membranes, advanced RO membranes should be developed to satisfy the low SEC requirement and high water quality. In addition, high-pressure resistant membranes should be developed to meet the needs of membrane brine concentrators.

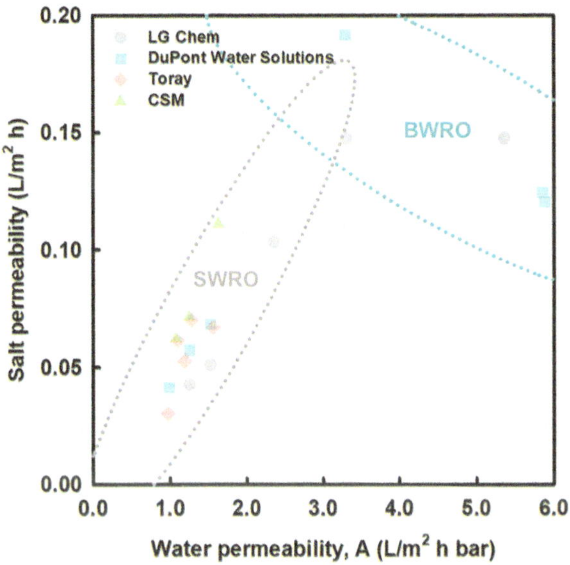

Figure 2.6 Water and salt permeability of SWRO and BWRO membranes.

2.6 RO CONFIGURATIONS

RO configurations determine the major performance of SWRO desalination plants. In this regard, RO configurations should be designed based on the targeting performance of SWRO desalination plants. Although various RO configurations have been adopted for different plants, the classifications of the RO process are not clearly summarized. In particular, the concept of pass and stage design has often been misunderstood in numerous studies. In addition, the overall RO system is explained only by a part of the RO configuration. Thus, RO configurations should be classified systemically.

The RO pass design should first be defined to present the overall RO configuration. When the RO system is composed of SWRO membranes, the RO system is classified as a single-pass RO system. The desalination process was performed using SWRO membranes only, and the permeate underwent post-treatment as a product. Furthermore, if the RO system is composed of both SWRO and BWRO membranes, the RO system is considered a two-pass RO. After the permeate is produced from the SWRO membranes, the BWRO membranes further desalinate the SWRO permeate. Depending on the features of the streamlines, a two-pass RO can be classified further.

After the RO pass design is defined, specific designs for the SWRO and BWRO systems can be classified (Figure 2.7). Most SWRO systems are composed of single-stage systems, but two-stage systems can be adopted for several plants. Further designs can be implemented to improve the performance of SWRO systems, such as pressure-center design and internally staged design. In contrast,

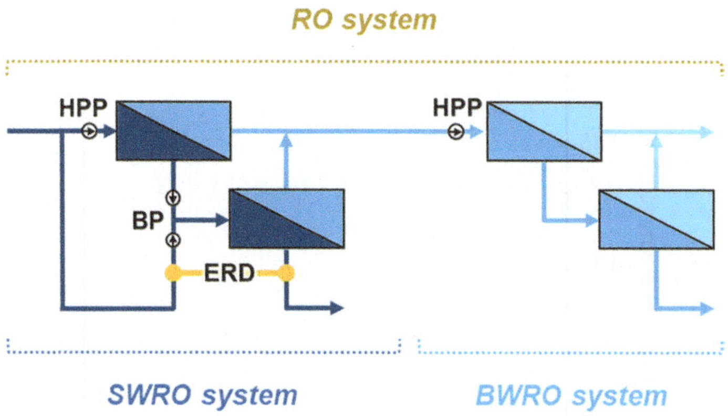

Figure 2.7 Composition of RO system. The SWRO system is a necessary part of the RO system, while the BWRO system is an optional process for improving water quality.

BWRO systems in SWRO desalination plants are generally configured as two-stage, and variations of two-stage BWRO, such as cascade, are widely adopted.

Owing to the application of multiple designs in a single SWRO desalination plant, it is necessary to define the type of RO system as (1) RO pass, (2) SWRO, and (3) BWRO configurations. However, the RO configuration is often referred to as the RO pass configuration. For example, single-pass RO refers to an RO system composed of single-stage SWRO, and two-pass RO refers to an RO system composed of single-stage and two-stage BWRO (Table 2.2).

2.6.1 RO pass configuration

The majority of RO pass designs are single- or two-pass RO for SWRO desalination plants. Single-pass and two-pass RO systems are also referred to as one- and double-pass RO systems, respectively. However, single- and two-pass RO systems are more general terminations for configurations. Various types of single- and two-pass RO systems have been developed and applied to satisfy water quality and energy efficiency.

Table 2.2 Summary of general termination for RO configurations.

Feed	Overall RO configuration	SWRO configuration	BWRO configuration
Seawater	Single pass	Single stage	N/A
	Two stage	Two stage	N/A
	Full two pass	Single stage	Two stage (or cascade)
	Partial second pass	Single stage	Two stage (or cascade)
	Split partial second pass	Single stage	Two stage (or cascade)

For overall RO configuration, it is preferred to name the overall pass configuration first (if the number of pass is multiple) and then specify stage configuration.

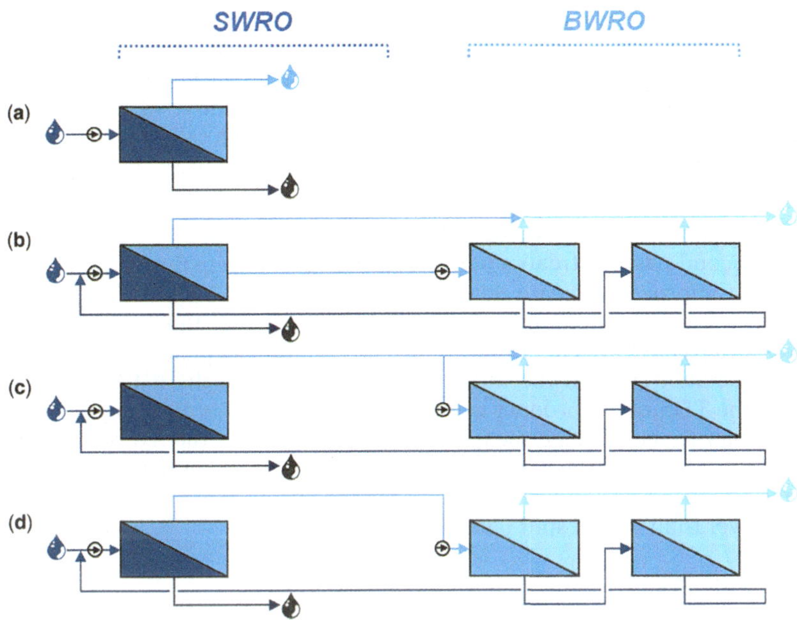

Figure 2.8 Scheme of RO pass configurations: (a) single-pass, (b) split partial second-pass, (c) partial second-pass, and (d) full two-pass RO.

2.6.1.1 Single-pass RO

Single-pass RO is only composed of an SWRO system without a BWRO system (Figure 2.8a). In single-pass RO systems, permeate is directly produced from seawater across SWRO membranes, and thus its TDS is relatively higher (e.g., 300–500 mg/L) than two-pass RO systems. Furthermore, the recovery of single-pass RO systems is higher than that of two-pass RO systems. This is because the SWRO permeate was not lost by the BWRO system. Owing to its simple configuration, single-pass RO can be widely used to produce freshwater for living in the environment and drinking. However, the TDS of the permeate would be higher in the region where high-salinity seawater is the source of feed (e.g., the Arabian Gulf). Thus, the application of a single-pass configuration is limited in terms of permeate quality.

2.6.1.2 Split partial second-pass RO (or split partial two-pass RO)

Split partial second-pass RO utilizes the front, and rear SWRO permeates separately to improve permeate quality with energy efficiency. When SWRO membranes are located in a PV system, the water flux and feed salinity are distributed differently. The feed salinity is relatively low, and the water flux is high in the front elements of the SWRO process. In contrast, the feed salinity is high, and the water flux is low at the rear elements. As a result, the permeate TDS is low for the front elements and high for the rear elements [17]. In this regard, it is thermodynamically energy-efficient to improve the final water quality by desalinating high-salinity permeate only and mixing it with low-salinity permeate. This concept is applied to the configuration of split partial second-pass (SPSP) RO. The SWRO front permeate (i.e., low salinity) is sent to a product tank directly, while the rear permeate (i.e., high salinity) is sent to the BWRO system for purer permeate production (Figure 2.8b).

2.6.1.3 Partial second-pass RO (or partial two-pass RO)

Partial second-pass RO treats a part of the SWRO permeate through the BWRO system to improve permeate quality (Figure 2.8c) [17]. The SWRO permeate is divided into two streams: one is sent to the product tank, and the other passes the BWRO system. Product quality can vary depending on the ratio of the partial streams. If the ratio of the stream sent to the BWRO system is higher, the TDS of the product is lower. Thus, the ratio is often higher during the summer to meet the permeate quality, as the TDS of the SWRO permeate is higher with an increase in temperature. However, as a part of the SWRO permeate is taken without splitting, the energy efficiency of partial second-pass RO is lower than that of split partial second-pass RO.

2.6.1.4 Full two-pass RO

Full two-pass RO wholly supplies the SWRO permeate to the BWRO without splitting the stream. Thus, the TDS of the SWRO permeate is significantly lowered by the BWRO system, and the TDS of the final product is low. In this regard, a full two-pass RO is employed when the permeate is highly pure (Figure 2.8d). Owing to its low TDS level, the product of full two-pass RO would not be adequate for drinking water production. Thus, the low-salinity product must serve as drinking water with remineralization or be utilized as industrial water demanding pure water. Though rarely, the permeate of full two-pass RO can be treated further by BWRO (i.e., full triple-pass RO) when extremely high-purity water is produced for industrial use [11].

2.6.2 SWRO system configuration

2.6.2.1 Single-stage SWRO

SWRO is generally composed of a single-stage configuration (Figure 2.9a). Multiple PVs are installed in an SWRO train, and the concentrate from the train is disposed of after its hydraulic energy is recovered. Because SWRO equipment and devices are standardized for single-stage SWRO operations, the configuration is a basic choice for SWRO design. However, the SWRO system configuration can be modified to overcome the limitations of the single-stage operation.

2.6.2.2 Two-stage SWRO

Two-stage SWRO can increase the recovery of the SWRO system by producing additional permeate from the second stage (Figure 2.9b). At the rear SWRO elements, the osmotic pressure of the feed almost reaches the hydraulic pressure of the feed. Because the net driving pressure (NDP) is low, a low amount of permeate is produced from the element. Thus, SWRO as a single stage is limited for high-recovery operations. However, when additional hydraulic pressure is

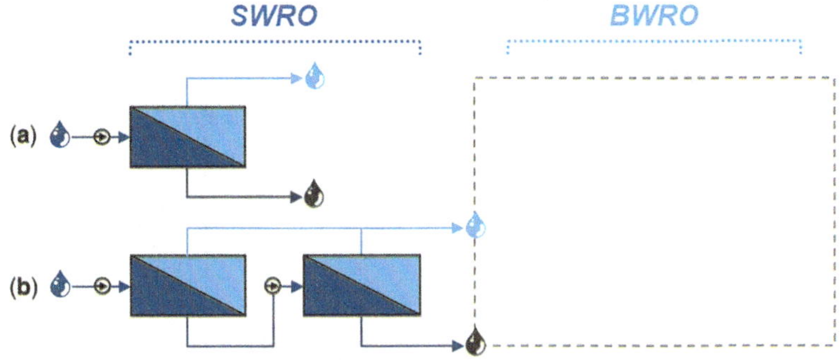

Figure 2.9 Scheme of SWRO system configurations: (a) single-stage and (b) two-stage SWRO.

applied to the feed, the permeate can be produced from the SWRO element because of the higher NDP. By elevating the hydraulic pressure through the inner boost pump between the first- and second-stage SWRO, the second-stage SWRO produces additional permeate. The ratio of permeate for the first and second stages is 2:1 as a rule of thumb [103, 104]. However, high-pressure-resistant SWRO membranes and equipment should be employed to overcome hydraulic pressures higher than 80 bar. As commercial SWRO membranes are designed to overcome a hydraulic pressure of 80 bar, a higher capital cost is required for the SWRO design. A two-stage SWRO design is often employed for retrofitting old plants to increase plant capacity.

2.6.2.3 Pressure center design

The energy efficiency of HPPs significantly affects the overall energy consumption of SWRO, as it is the dominant energy-consuming unit. To achieve low-energy consumption in RO systems, high-efficiency HPPs should be employed in the systems. Notably, the mechanical efficiency of the pump is highly associated with its capacity, and the efficiency is improved when the capacity of the HPP is increased. Thus, a pressure-center (or three-center) design is developed to lower the energy consumption of the RO system by increasing the capacity of HPPs and BPs. In particular, the pressure-center design combines several SWRO trains to the main feed line to supply pressurized feed simultaneously using larger-sized HPPs (Figure 2.10). In addition, the design allows the SWRO system to produce varying amounts of permeate more effectively by integrating multiple SWRO trains. This leads to a more flexible operation of the RO system in accordance with the water demand. In contrast, the combination of multiple SWRO trains was not beneficial in terms of the maintenance of the RO system. While each SWRO train can be repaired separately in a conventional system, all SWRO trains must be stopped to repair a single SWRO train. Despite its advantages in SEC reduction and disadvantages in maintenance, several SWRO desalination plants have adopted the design of SWRO systems to reduce SEC. Because the design was developed by an Israeli company, IDE Technologies, the design has been mainly applied to Israeli plants [11, 105–107].

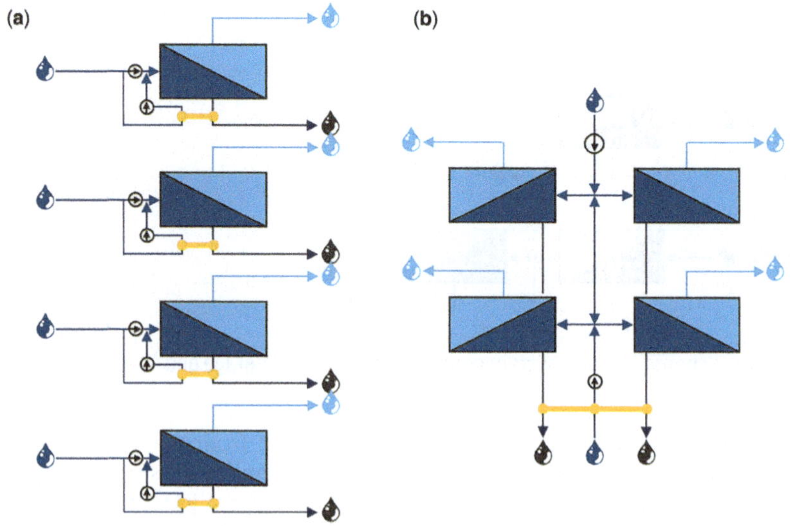

Figure 2.10 Scheme of pressure-center design: (a) typical RO trains and (b) RO trains with pressure-center design.

2.6.2.4 Internally staged design

Internally staged design (ISD) utilizes different RO membranes in a PV to acquire performance benefits (Figure 2.11) [17, 108]. When RO membranes are placed in a PV, high-rejection membranes and high-flux membranes are in the front and rear, respectively. The arrangement exhibits several operational benefits compared to RO systems employing single-type membranes.

ISD can reduce the fouling propensity by decreasing the water flux at the front elements. Front RO elements are easily fouled with organic and colloidal particles owing to high-flux operation. However, when the front RO membranes are high-rejection membranes (i.e., low water permeability), the membranes are operated at lower fluxes. In other words, fouling is mitigated by ISD.

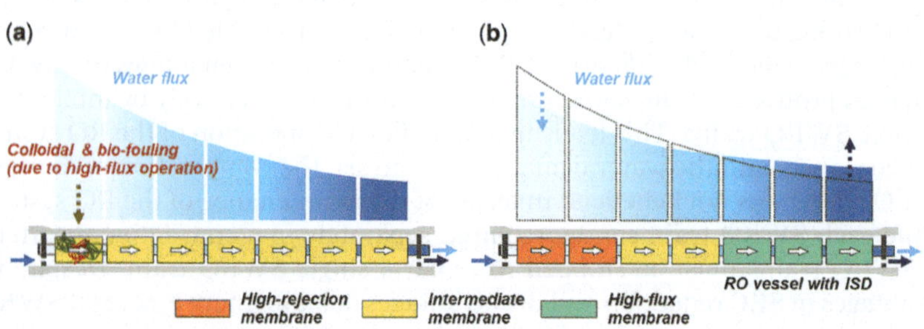

Figure 2.11 Scheme of internally staged design (ISD) in RO process: (a) normal design and (b) ISD.

More diverse RO performance can be obtained by utilizing different types of membranes. Achievable RO performance is limited when single-type membranes are used. In addition, RO systems equipped with different membranes can exhibit wider RO performance. The use of an optimized membrane arrangement allows the RO system to operate with higher energy efficiency under the control of operating conditions.

With the operational benefits, it has been reported that Las Palmas III, Mazarrón, and El Coloso SWRO plants adopted ISD to SWRO systems [109]. However, the complexity of maintenance still needs to be addressed to utilize ISD [108].

2.6.3 BWRO system configuration

2.6.3.1 Single-stage BWRO

Brackish water has a TDS of 500–15 000 mg/L. However, BWRO plants typically treat feeds of 500–10 000 mg/L [102]. Because there are various sources of brackish water, it can be classified with more detailed criteria: water with a TDS of 500–2500 mg/L is classified as low saline and that with a TDS of 2500–10 000 mg/L as high saline. Notably, the osmotic pressure of brackish water is not as high as that of seawater and is still lower than that of seawater even when it is concentrated up to 90% in low-salinity brackish water. Because of the low osmotic pressure, BWRO systems can be operated at a low hydraulic pressure, and the system recovery can be increased substantially. When the feed is low-salinity brackish water, a single-stage (or -pass) system can be adopted (Figure 2.12a) [102]. However, high recovery cannot be achieved through single-stage/pass BWRO systems, even though brackish feed has low osmotic pressure.

2.6.3.2 Two-stage BWRO

Two/multistage systems are common in BWRO applications to achieve high recovery. The target recovery is 70–90% with a water flux of 20–40 L/m^2 h, depending on feed salinity and characteristics [17], and the value is still higher than that of single-stage (-or pass) operation. The staged system generally follows a 2:1 array for a two-stage configuration (where the ratio of the PV number for the first and second stages is 2:1) and 3:2:1 for the three-stage configuration [102].

However, BWRO systems differ depending on the salinity of brackish water. For low-salinity brackish water, two/multistage BWRO systems are commonly used as well as single-stage BWRO systems, and no BPs are installed between the first and second stages (Figure 2.12b). This is because the hydraulic pressure is sufficiently high to overcome the osmotic pressure of the feed. In contrast, when the feed is high-salinity brackish water, two-stage BWRO equipped with BPs between the stages is applied to further produce permeate from the second stage (Figure 2.12c).

Two/multistage BWRO systems can also be implemented in SWRO desalination plants to improve the water quality. In two-pass SWRO systems, the first pass comprises SWRO membranes and the second pass of BWRO membranes [11, 17]. As the feed for the second pass is a low-salinity brackish, it is usually configured as a two-stage BWRO system without inner BPs.

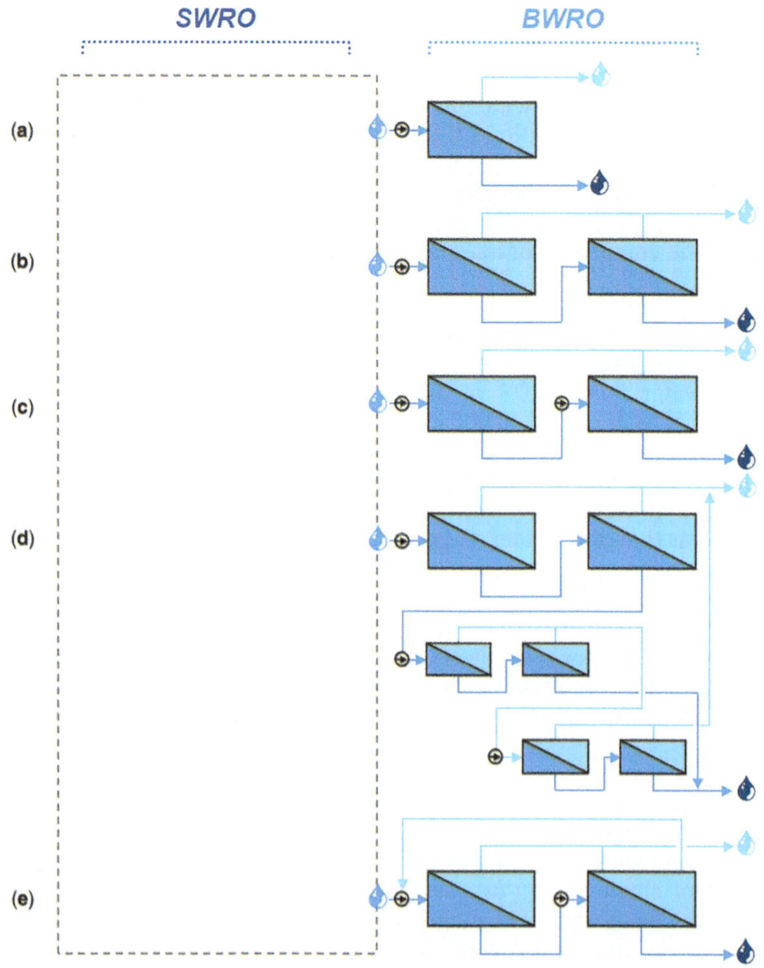

Figure 2.12 Scheme of BWRO configurations: (a) single-stage, (b) two-stage (low-salinity water), (c) two-stage (high-salinity water), (d) multistage/cascade (low-salinity water), and (e) permeate circulation process (PCP).

2.6.3.3 Cascade BWRO

To achieve higher water quantity and quality, several SWRO desalination plants use a cascade BWRO system similar to multistage systems (Figure 2.12d). The cascade system can be varied, but it is usually a set of BWRO systems where the first-pass permeate is treated by a BWRO multistage system (2–4 stages) to increase water recovery. Furthermore, the permeate produced from the rear BWRO stages is further treated by another set of BWRO stages to reduce TDS.

2.6.3.4 Permeate circulation BWRO

In contrast, a permeate circulation process (PCP) has been applied to BWRO systems (i.e., second pass) of Shuqaiq II SWRO desalination plants where the BWRO rear permeate is circulated back to the feed to improve water quality (Figure 2.12e) [11, 110]. Likewise, BWRO integrated with SWRO has been extensively adopted, even in SWRO desalination plants.

BWRO systems can be utilized for both treating brackish water and improving the water quality of SWRO systems. In other words, the role of BWRO will increase in the future desalination market, regardless of feed salinity. Because the main benefit of BWRO is a high-recovery operation, BWRO systems should be developed further to maximize their recovery while ensuring permeate quality.

2.7 POST-TREATMENT

The permeate of SWRO membranes is usually depleted in minerals and thus has a high corrosion potential. This makes SWRO permeate aggressive to the components of water distribution systems such as pipes, pumps, and tanks (when tankers distribute water). The aggressiveness of the SWRO permeate is controlled by increasing its hardness and alkalinity. However, in some cases where the intake is from the subsurface, there might be a need to remove carbon dioxide and hydrogen sulfide because subsurface intake water is high in carbon dioxide and hydrogen sulfide, which are not removed during the SWRO process.

In some SWRO plants producing potable water for industrial and domestic purposes, a portion of the pre-treated feed is blended with the SWRO permeate for many reasons, such as water stability, corrosion control, capital operating expenditure, and footprint reduction. The blending proportion is dependent on both the treated water quality and pretreated water quality. Blending can also be done with water from other sources, such as groundwater and potable water. Overall, blending improves the stability and taste/dietary components of SWRO-treated water. Blending seems to be an easy post-treatment strategy to overcome most of the challenges posed by SWRO permeates; however, there are concerns regarding the potential addition of anions such as bromide and iodide, which could cause disinfection byproducts in the treated water. Therefore, proper assessment of both the pretreated and post-treated water is essential to determine the ratio of blending for water quality control. The pretreated water to be blended with SWRO permeate must also be evaluated for microbial and chemical contaminants and duly removed before blending with SWRO permeate as they have the potential to compromise the safety of the treated water [111]. Blending should not allow the introduction of pathogenic microorganisms into the treated water. In the case where potable water from other sources is used to blend SWRO permeate, caution should be exercised in selecting piping materials to eliminate the possible leaching of contaminants from pipe materials [112]. Generally, blending is not sufficient for SWRO permeate stability; thus, the alkalinity and hardness of the water need to be increased. Nevertheless, appropriate blending reduces the quantity of chemicals required for corrosion control and stability.

SWRO plants have diverse post-treatment strategies depending on the use of the permeate as regulated by the authorities. However, conventional post-treatment steps in the SWRO process include stabilization and corrosion control, disinfection, and air stripping and degasification (especially of CO_2 and H_2S gases).

2.7.1 Stabilization and corrosion control

Permeate stabilization is believed to be one of the most essential aspects of the SWRO desalination process. Corrosion prevention would significantly reduce the frequency of pipes, pumps, and other distribution unit replacements and, in turn, reduce the operating costs of the process. However, stabilizing SWRO permeate could be demanding, as the chemical and dosage requirements differ from plant to plant. Therefore, each plant needs to be evaluated to assess its control needs when designing a permeate stabilization strategy. Over the years, SWRO plant operators have used three parameters to improve the stability of the SWRO permeate. These parameters included pH, alkalinity, and calcium carbonate adjustment. Each parameter has its own characteristics that can help improve the stability of the permeate within the process and during distribution.

2.7.2 pH adjustment

The pH of the SWRO permeate is mostly a factor of bicarbonate alkalinity and other elements such as calcium, sulfate, chloride, DO, total dissolved solids, and boron content. The pH of the permeate can be adjusted using sodium hydroxide, potassium hydroxide, carbon dioxide, lime, or soda ash. The pH of the permeate affects the metal ion dissolution potential and the precipitation of insoluble compounds. This means that metal material pipes can be prevented from corrosion by lowering their dissolution. The precipitation of insoluble carbonate on the surface of the pipes could serve as a coating agent to prevent corrosion. Insoluble carbonate was precipitated at an elevated pH (>8.4). The possibility of biofilm formation in the distribution pipes is lower if the permeate pH is adjusted to ≥ 9.0. pH adjustment may not be suitable for permeates with high hardness (>150 mg/L $CaCO_3$). Inappropriate pH could cause problems such as copper pitting, trihalomethane production ($pH > 8.1$), disinfection byproducts ($pH > 7.8$), $CaCO_3$ sealing ($pH > 7.9$), and growth of ammonia-oxidizing bacteria ($pH < 8$).

2.7.3 Alkalinity adjustment

The buffering capacity of the SWRO permeate is measured by its alkalinity. The corrosion tendency and susceptibility to pH changes by SWRO permeate are reduced at elevated alkalinity. Therefore, the alkalinity of the water is adjusted so that it can induce the precipitation of insoluble compounds on the surface of the water distribution units. The coating of distribution surfaces could help prevent corrosion and subsequently increase the lifespan of such distribution components. The alkalinity of the permeate is directly linked to the pH of water, bicarbonate, carbonate, and hydroxide ions. Alkaline water can easily produce H^+ and OH^- ions to neutralize the effect of pH change, thus stabilizing the water. Although alkalinity adjustment improves the buffering capacity of the permeate, there are still some challenges of operation and maintenance costs and high carbonate scaling on pipelines.

2.7.4 Hardness ($CaCO_3$) adjustment

Hardness (CaCO3) adjustment is often done to control the corrosion ability of the permeate. The goal of this approach is to develop a $CaCO_3$ film on the surface of the distribution unit. Usually, pH and alkalinity adjustment create an environment suitable for the precipitation of Ca and CO_3. The precipitation of these compounds is achieved by introducing CO_2, $CaCO_3$, soda ash, and lime into the water. The major challenge of this approach is that $CaCO_3$ films are not formed on the pipes alone but also in all the distribution units, such as pipes, pumps, and boilers. Similarly, this post-treatment strategy could also lead to sulfate precipitation.

2.7.5 Disinfection

The SWRO-treated water is temporarily stored before distribution. The treated water experiences microbial contamination. Therefore, disinfection is conducted to deactivate residual microbes in the treated water and protect it from subsequent microbial contamination during storage and distribution. The selection of disinfection chemicals depends on costs, safety, and availability [113]. Post-treatment disinfection is achieved with liquid or gas chlorine, production of hypochlorite on-site, calcium hypochlorite, and bulk hypochlorite. Chlorine disinfection is mostly used in SWRO processes. If the SWRO permeate is not contaminated with organic matter during remineralization, disinfection byproducts are not generated during the disinfection process. However, if the treated water is contaminated with disinfection byproduct precursors such as dissolved organics, chlorination would result in the generation of disinfection byproducts. Similarly, chlorine disinfection could also help in the removal of H_2S because of the ability of chlorine to react selectively with sulfides. Chlorine dosages between 5 and 10 mg/L is sufficient to preserve the water from microbial attack.

2.7.6 Aeration and degasification

When SWRO intake is from the subsurface or well, it is characterized by low DO and high H_2S. DO and H_2S are not eliminated during the desalination process, thus making their way to the permeate. Therefore, in this case, the permeate must be aerated and degassed before distribution. Aeration is performed to increase the DO content of the permeate, remove volatile organic compound contaminants, degas H_2S, and remove CO_2, consequently reducing the corrosion ability and increasing pH. The SWRO permeate is aerated either by allowing the permeate to fall through air or by injecting air into the permeate.

2.8 DISCHARGE

All SWRO plants, irrespective of their size, need to discharge their waste/concentrate. The discharge from the SWRO plant is usually from different sections of the SWRO process. These sections include waste from pretreatment units such as sludge (from sedimentation tank and DAF), backwash water, concentrate from membrane units (UF, NF, and RO), metals from corrosion and waste from processes, and membrane cleaning [114]. This discharge is characterized by physical properties (salinity, 65 000–85 000 ppm, temperature, 5-10°C of ambient seawater temperature, suspended solids and coagulants, antiscalants, metals, and cleaning chemicals [61, 115].

2.8.1 Conventional discharge strategy

The aim of conventional discharge is to release SWRO brine through an open pipe directly into the ocean deep without any detrimental impact on marine life. Because the density of seawater is a function of its salinity, SWRO brine tends to sink and dilute slowly on the seabed. Therefore, discharge pipes are often placed in locations where the discharge can be easily and quickly mixed with the receiving seawater. Sometimes, the pipes are equipped with diffusers (nozzles used to increase the mixing of the concentrate to prevent stagnation on the ocean floor). Alternatively, the effluent from the SWRO plant is mixed with the treated discharge from the sewage plant to dilute the brine. However, owing to the environmental impact of direct discharge into the ocean, SWRO effluents are generally treated before discharge to meet the discharge regulations of the region.

When designing or deciding the proper discharge site in the open ocean discharge, it is important that the concentrate is not discharged in locations with stressed and endangered marine ecosystems. It is equally important to discharge concentrate in locations where there is a strong underwater current for the accelerated mixing of the concentrate.

2.8.2 Discharge to injection wells

Injection well systems are considered environmentally benign discharge methods, and an injection zone is available that can accept the discharge without any significant perturbation to marine life. Injection wells are of two types: shallow and deep wells. The deep-well system is usually hundreds of meters below the land surface, and the discharge is expected to remain underground permanently. Generally, the injection well discharge method is characterized by injecting SWRO waste into an underground aquifer isolated from other water aquifers. It is important that the aquifer has the capacity to collect such waste through the life of the SWRO plant (usually 25–30 years) [116].

2.8.3 Discharge to offshore galleries and trenches

The SWRO concentrate and waste are discharged into the infiltration trenches. Infiltration trenches are mostly perforated pipes buried parallel to the beach. This discharge method is used in small-scale desalination plants. One advantage of infiltration trenches is their ability to slowly diffuse SWRO brine offshore without any significant impact on marine life. Large-scale SWRO desalination plants can use the beach gallery method because the brine can be discharged via the top of the gallery, unlike in infiltration trenches, where pipes are run horizontally along shallow sediments.

2.8.4 Zero liquid discharge (ZLD)

Owing to increasing concerns of indiscriminate discharge and its impact on the environment, regulatory bodies are implementing stricter discharge requirements that are often difficult to meet by conventional discharge methods [117]. Therefore, it is a matter of urgency to use discharge strategies that will satisfy the requirement for discharge and, perhaps during the process, improve the performance and the efficiency of the entire process. ZLD is considered one of the safest discharge methods, without any significant adverse effects on the environment. This method is beneficial for SWRO plants located in inland regions. ZLD can be achieved using evaporation ponds, crystallizers, and spray dryers.

In the evaporation pond discharge method, the SWRO effluent collected in an impervious lined shallow earthen basin can slowly evaporate by utilizing energy from the sun. The impervious lining is essential to prevent waste from contaminating underground aquifers. After the water is completely evaporated, the remaining solid waste is collected and disposed of. The major drawback of this approach is the requirement for a large expense of land, which could significantly increase the plant's capital costs. This method is also limited to dry and semi-arid regions.

Crystallizers are cylindrical vessels that can be heated by either a steam source or a vapor compressor to produce a crystal/precipitate. Crystallizers can be used to achieve ZLD in SWRO discharge without the need for frequent cleaning and excessive scale development.

Spray dryers are an alternative to crystallizers for producing salt crystals from SWRO brine. The advantage of spray dryers over crystallizers is that in spray dryers, the product shape, size distribution, and density can be controlled [118].

2.8.5 Dilution of concentrate using forward osmosis process

The forward osmosis (FO) process is a promising SWRO concentrate diluting technology owing to its ability to use an osmotic pressure gradient to draw water. In the FO process, a high saline SWRO concentrate is used to draw water from the domestic wastewater and in so doing, the SWRO concentrate is diluted by the automatically treated wastewater. Although this approach has not yet been commercialized for discharge, it is a promising discharge method suitable for meeting the discharge requirements.

doi: 10.2166/9781789061215_0037

Chapter 3

Energy consumption in SWRO operation

3.1 MEMBRANE TRANSPORT MECHANISM IN A SMALL-SCALE SYSTEM

3.1.1 Solution–diffusion theory

Porous membranes such as ultrafiltration (UF) or microfiltration (MF) have larger pore sizes than the size of the solute molecules. Therefore, the transport mechanism of the porous membrane is dominated by size exclusion or sieving. The pore flow model is typically used to describe the transport mechanism in UF and MF membrane systems [119, 120]. The separation performance of UF and MF systems can be estimated using the pore flow model.

Unlike porous membranes, the transport of salt and water molecules in a dense membrane such as an RO membrane is described by the solution–diffusion model. In the solution–diffusion model, the transport mechanism is determined by dissolution in the dense membrane material and diffusion through the membrane [121]. The separation in the dense membrane is governed by (1) the solubility differences of the salt and water in the membrane, and (2) the difference in the diffusion rates of the salt and water through the membrane. The solution–diffusion model is shown in Figure 3.1.

Thus, the permeability through the membrane is calculated based on the solubility and diffusivity. The general expression for the flux in the solution–diffusion model is given as follows [121]:

$$J_i = -Y_i \frac{\mathrm{d}\mu_i}{\mathrm{d}x}, \tag{3.1}$$

where μ is the chemical potential of component i, and Y is the coefficient of proportionality linking the chemical potential driving force to the flux.

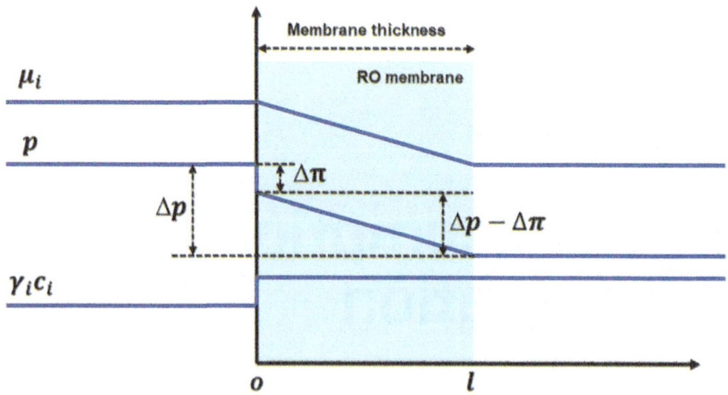

Figure 3.1 Schematic diagram of the solution–diffusion model in an RO membrane.

The differential term of the chemical potential is expressed in terms of the concentration gradient and pressure gradient, as follows:

$$d\mu_i = RT d\ln(\gamma_i c_i) + v_i dp, \tag{3.2}$$

where c_i is the molar concentration of component i, γ_i is the activity coefficient, v_i is the molar volume of component i, R is the gas constant, T is the temperature, and p is the pressure. Then, the flux term is expressed as follows:

$$J_i = -\frac{RTY_i}{c_i}\frac{dc_i}{dx} = -D_i\frac{dc_i}{dx}. \tag{3.3}$$

RTY_i/c_i can be replaced by the diffusion coefficient D_i. By integrating Eq. (3.3) over the membrane thickness (l), the flux term is obtained as follows [121]:

$$J_i = \frac{D_i(c_{io(m)} - c_{il(m)})}{l}, \tag{3.4}$$

where $c_{io(m)}$ and $c_{il(m)}$ are the concentrations of component i at positions o and l in the RO membrane, respectively. $c_{io(m)}$ is expressed using the sorption coefficient, K_i, as follows:

$$c_{io(m)} = K_i c_{io}, \tag{3.5}$$

$$c_{il(m)} = K_i c_{il}. \tag{3.6}$$

By considering the effect of pressure on the chemical potential in an incompressible fluid, $c_{il(m)}$ can be expressed as follows [121]:

$$c_{il(m)} = K_i c_{il} \exp\left(\frac{-v_i(p_o - p_l)}{RT}\right). \tag{3.7}$$

Substituting Eqs. (3.5), (3.6), and (3.7) into Eq. (3.4) yields the following:

$$J_i = \frac{D_i K_i}{l}\left[c_{io} - c_{il}\exp\left(\frac{-v_i(p_o - p_l)}{RT}\right)\right]. \tag{3.8}$$

At osmotic equilibrium, the water flux is zero. Thus, the correlation between c_{io} and c_{il} can be obtained as follows:

$$c_{il} = c_{io} \exp\left(\frac{v_i \Delta\pi}{RT}\right). \tag{3.9}$$

Then, Eq. (3.8) can be arranged as follows:

$$J_i = \frac{D_i K_i c_{io}}{l}\left[1 - \exp\left(\frac{-v_i\left(\Delta p - \Delta\pi\right)}{RT}\right)\right], \tag{3.10}$$

where ΔP is the pressure difference across the RO membrane (p_0 - p_l). Eq. (3.10) can be rewritten as follows based on the approximation that 1 - exp(x) is close to x:

$$J_i = \frac{D_i K_i c_{io} v_i (\Delta p - \Delta\pi)}{lRT}, \tag{3.11}$$

This equation can be simplified as follows:

$$J_i = A(\Delta p - \Delta\pi), \tag{3.12}$$

$$A = \frac{D_i K_i c_{io} v_i}{lRT}, \tag{3.13}$$

where A is a permeability constant equal to the term $D_i K_i C_{io} v_i / lRT$.

3.1.2 Permeate resistance

As described in the previous section, the water permeation mechanism through the RO membrane can be expressed in a simplified manner based on water permeability. The salt flux can also be described using the same mechanism but utilizing only the concentration difference. The flux equations are usually expressed based on the driving force and membrane permeability as follows [122, 123]:

$$J_w = A(\Delta P - \Delta\pi), \tag{3.14}$$

$$J_s = B(C_m - C_p), \tag{3.15}$$

where J_w is the water flux, A is the water permeability of the RO membrane, ΔP is the applied pressure on the feed solution, $\Delta\pi$ is the osmotic pressure difference between the feed solution and the permeate, J_s is the salt permeability, B is the salt permeability of the RO membrane, C is the concentration, and the subscripts m and p denote the membrane surface on the feed solution side and the permeate, respectively. It should be noted that the concentration at the surface of the membrane is significant for estimating the water and salt fluxes precisely. Concentration polarization should be considered to estimate the surface concentration of the membrane, which will be discussed in the next section.

The water and salt permeabilities are estimated from the intrinsic membrane transport properties (P_w for water permeation and P_s for salt permeation) as follows [124]:

$$A = \frac{P_W}{L}\frac{M_W}{RT}, \tag{3.16}$$

$$B = \frac{P_s}{L}, \tag{3.17}$$

where L is the membrane thickness, M_w is the molecular weight, R is the gas constant, and T is the absolute temperature. A and B have a trade-off relationship, and thus the selectivity of the RO membrane can be estimated based on the empirical correlation between the values of A and B as follows [125]:

$$B = \frac{L^{\beta}}{\lambda}\left(\frac{RT}{M_w}\right)^{\beta+1} A^{\beta+1}, \tag{3.18}$$

where β and λ are the correlation parameters, which are determined empirically.

3.1.3 Concentration polarization

Based on the equations derived from the solution–diffusion model, the water and salt permeation in the RO system seem to be linearly correlated with the driving force. However, nonlinear relationships of the water and salt permeation are found in actual RO systems with increased driving force. The main reason for this nonlinear behavior is the concentration polarization on the surface of the membrane. Because a semi-permeable membrane has high selectivity for salt permeation, most of the salt molecules are retained at the surface of the membrane, while the water molecules can permeate through the membrane. Thus, the concentration at the surface of the membrane is polarized, and the surface concentration is higher than the bulk concentration. This means that the effective osmotic pressure should be estimated based on the surface concentration rather than the bulk concentration. If the concentration polarization is not considered, the effective osmotic pressure in the RO system will be underestimated. As a result, it will not be possible to calculate the water flux and energy consumption in the RO system precisely. Therefore, the consideration of concentration polarization is very important in membrane desalination systems such as RO.

Measuring the concentration at the membrane surface is not practical. Therefore, the estimation of the surface concentration is conducted using a modeling approach. The modeling of concentration polarization is usually derived based on the boundary layer film theory [126].

A detailed schematic diagram is shown in Figure 3.2. Three salt flux terms at the surface of the RO membrane should be considered in the concentration polarization modeling: the diffusive flux, convective flux, and salt flux [127]. Assuming a steady state, these flux terms can be arranged as follows [128]:

$$D\frac{\mathrm{d}C(x)}{\mathrm{d}x} + J_w C(x) = J_w C_p, \tag{3.19}$$

where D is the diffusion coefficient. The first and second terms on the left-hand side of this equation are the diffusive flux and convective flux, respectively, and the term on the right-hand side is the salt flux. In the boundary layer film theory, a concentration gradient between the bulk concentration and surface concentration appears along the boundary layer thickness (δ). Therefore, the

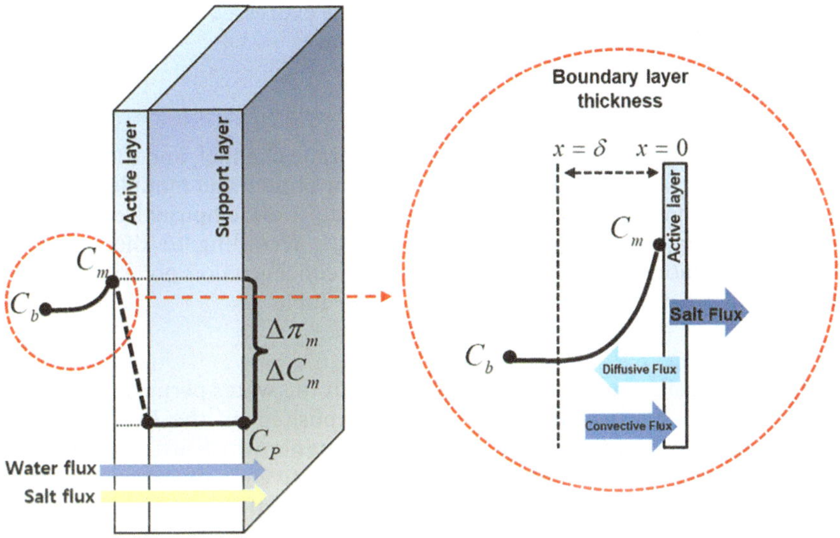

Figure 3.2 Schematic diagram of concentration polarization and the detailed salt fluxes on the surface of the RO membrane.

boundary conditions of this differential equation can be defined as $C = C_b$ at $x = \delta$ and $C = C_m$ at $x = 0$. Then, Eq. (3.19) can be solved as follows:

$$\exp\left(\frac{J_w\delta}{D}\right) = \frac{C_m - C_p}{C_F - C_p}. \tag{3.20}$$

In the boundary layer film theory, the ratio D/δ can be replaced by the mass transfer coefficient (k) in each film [129]. The concentration polarization model can be derived as a function of the water flux, mass transfer coefficient, bulk concentration, and permeate concentration.

$$\exp\left(\frac{J_w}{k}\right) = \frac{C_m - C_p}{C_F - C_p} \tag{3.21}$$

It is difficult to estimate the mass transfer coefficient theoretically. In very limited cases, an analytical solution for the mass transfer coefficient can be obtained. For example, assuming that the laminar boundary layer is formed and the Schmidt number is unity, the Blasius solution can be used to obtain the mass transfer coefficient [130]. However, the practical situations are not simple. In an RO membrane module, a spacer is usually included to create turbulent flow inside the membrane channel because turbulent flow is helpful for reducing the effect of concentration polarization inside the membrane channel. Thus, the flow regime cannot be limited to the laminar flow region. To overcome these limitations, the empirical correlation of the Sherwood number is widely utilized [128, 131]. The Sherwood number can be expressed as a function of the Reynolds number and Schmidt number, as follows:

$$Sh = \frac{kd_h}{D} = aRe^b Sc^c, \tag{3.22}$$

where d_h is the hydraulic diameter of the membrane channel, and the empirical coefficients a, b, and c are dependent on the flow regime and module structure. Many studies have been conducted to estimate these empirical coefficients. By selecting appropriate empirical coefficients according to the operating conditions, a more effective estimation of the concentration polarization can be realized. A list of coefficients is presented in Table 3.1.

3.1.4 RO fouling

Fouling and scaling are usually correlated with the water permeability of the RO membrane. The foulants and scalants deposited on the RO membrane hinder water permeation across the RO membrane, which usually increases the water permeability. Thus, RO fouling negatively influences the SEC. Although a general estimation methodology for the effect of the extent of RO fouling on the water permeability is not easily obtained, these effects should be considered and applied to the water permeability of the RO membrane. The water permeability of a virgin RO membrane is only applicable at the start of the RO operation.

Table 3.1 Sherwood number correlation coefficients and validated conditions.

Module type	Sherwood number correlation	Validated conditions	Ref.
Flat sheet	$Sh = 1.62\left(Re \cdot Sc \cdot \frac{d_h}{L}\right)^{0.33}$	Laminar flow ($Re < 2100$)	[134]
	$Sh = 0.34Re^{0.75}Sc^{0.33}$	$10^4 < Re < 10^5$	[135]
	$Sh = 0.023Re^{0.8}Sc^{0.33}$	$Re > 10^5$	[135]
	$Sh = 0.2Re^{0.57}Sc^{0.4}$	$Re < 50$	[132]
	$Sh = 1.964Re^{0.406}Sc^{0.25}$	$0.7 < Re < 1.7$	[136]
	$Sh = 0.023Re^{0.875}Sc^{0.25}$	$300 < Sc < 700$	[137]
	$Sh = 0.0149Re^{0.88}Sc^{0.33}$	$Sc > 100$	[138]
	$Sh = 0.107Re^{0.9}Sc^{0.5}$	$0.5 < Sc < 10$	[139]
Spiral-wound	$Sh = 0.023Re^{0.875}Sc^{0.25}$	$Re > 2100$	[140]
	$Sh = 0.2Re^{0.57}Sc^{0.40}$	Common commercial spacer	[132]
	$Sh = 0.16Re^{0.605}Sc^{0.42}$	$L/D = 8$, spacer angle $= 90°$	[132]
Hollow fiber	$Sh = 0.2Re^{0.6}Sc^{0.33}$	$40 < Re < 1000$	[141]
	$Sh = 0.17Re^{0.6}Sc^{0.33}$	$20 < Re < 200$	[142]

Source: Koutsou *et al.* [132] and Shibuya *et al.* [133].

3.1.4.1 Particulate/colloidal fouling

Particulate or colloidal fouling refers to membrane fouling caused by the deposition of particles or colloids on the membrane surface. Generally, these types of foulants are organic and inorganic particles suspended in the source water, which includes a wide variety of matter such as debris, silt, colloidal silica, iron, aluminum, polysaccharides, and natural organic matter (NOM). These solids exist in an insoluble form. Particulate and colloidal foulants are usually grouped together, but they can be classified according to their size. Particulate matter is defined as particles with sizes of greater than 1 μm, and colloidal matter comprises particles within the size range of 1–0.001 μm (Table 3.2) [143].

Because particulate compounds are relatively large, they are the easiest foulants to remove. These solids cannot pass through the RO membrane and are completely retained and concentrated on the feed side. As a result, they form a thick cake layer, which degrades the process performance. However, most of these foulants can be removed by implementing well-designed conventional pretreatment processes such as flocculation/sedimentation and granular media filtration prior to the RO step.

Colloidal foulants are smaller than their particulate counterparts. These foulants can generally be further classified into organic and inorganic foulants. Organic colloidal foulants include proteins, hydrocarbons, polysaccharides, and NOM, whereas inorganic colloidal foulants include colloidal silica, iron, aluminum, and manganese [144]. Although colloids do not usually exist in large coagulated forms like particulate matter, these colloids can be present in undissolved forms in the source water in the RO step. When colloidal particles are concentrated through separation by the RO membrane, they join together to form a sediment and cake layer on the surface of the membrane [145].

In addition to considering the effect of physical parameters such as size, the chemical interaction between the foulants, membrane, and hydrodynamic conditions of operation should be considered as important variables for particulate and colloidal fouling. Both foulants are known to carry negative surface charges in the pH range of natural seawater, in which case RO membranes with similar affinities are less susceptible to fouling. Likewise, highly hydrophilic and less rough RO membranes are less susceptible to the same problem [146]. Hydrodynamic conditions such as the crossflow velocity and operating pressure can also significantly affect the propensity for fouling. The higher the crossflow rate is, the more difficult it will be for foulants to settle on the membrane surface, and thus they will be removed with the retentate stream. When a high-pressure RO system is in operation, the applied pressure pushes the foulants toward the membrane, thus hindering back diffusion as well as inducing compression of the existing fouling cake layer.

Table 3.2 Size and composition of particulate and colloidal matter.

Foulant	Approximate size	Composition
Particulate matter	>1 μm	Debris, clay, silt
Colloidal matter	1–0.001 μm	Colloidal silica, iron, polymers, plankton

3.1.4.2 Organic fouling

Organic fouling is caused by organic substances dissolved within the feed solution, which may be naturally occurring or artificial compounds. Some common organic materials include humic acid (HA), bovine serum albumin (BSA), sodium alginate (SA), algae, polysaccharides, transparent exopolymer particles (TEPs), antifoaming agents, and polyacrylic polymers [147]. Similar to particulate and colloidal foulants, these organic molecules and microorganisms are easily rejected by RO membranes and conventional pretreatment processes because of their relatively large size. However, difficulties in operation can occur in the RO step and pretreatment processes because of the sticky characteristic of organic foulants. This characteristic allows the organic compounds to be adsorbed onto the membrane surface (Figure 3.3). The deposition of particulate or colloidal matter is quite different from that of the cake layer. The sediment of organic foulants forms a gel layer, and microorganisms are easily deposited on the gel layer to form a biofilm [148].

It is important to understand the foulant–foulant and foulant–membrane interactions to elucidate the mechanism of organic fouling in detail. The initial interaction between the bulk foulants and membrane surface causes the organic foulants to be deposited on the surface of the RO membrane. The extent and rate of gel layer formation are determined by the hydrophilicity and surface charge between the organic compounds and the RO membrane. When the gel layer on the membrane surface is covered to some extent, the intermolecular adhesion between the bulk foulants and the foulants on the membrane, that is, the foulant–foulant interaction, becomes the dominant force. This foulant–foulant interaction is the main factor in estimating the thickness and compactness of the gel layer. The gel layer formed at the surface of the RO membrane increases the hydraulic resistance, which impedes the water permeability of the RO membrane [149].

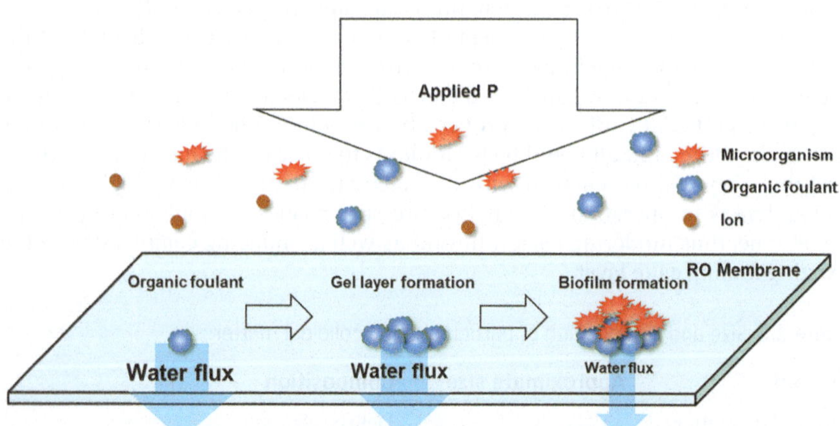

Figure 3.3 Membrane fouling by organic foulants.

Organic fouling is mainly attributed to TEPs. These foulants are organic compounds that are released by microorganisms, and they mainly consist of polysaccharides and amino sugars. In open seawater, which is utilized as typical feed water in the SWRO process, biodegradable organic matter, commonly known as algal organic matter (AOM), is excreted by marine algae during algal blooming periods, and TEPs comprise the high molecular weight fractions of those AOM compounds. TEPs serve as a nutritional platform for bacteria and other microbes to thrive owing to the extremely adhesive characteristics of the TEP. Furthermore, the same adhesive characteristics of the organic fouling layer and TEPs hinder effective washing or backwashing of the fouled membrane surface. It is thus crucial to operate adequate pretreatment processes before the RO process to minimize damage to the RO membrane.

3.1.4.3 Biofouling

Biofouling is regarded as one of the greatest challenges in RO processes. Biofouling is formed in the stage subsequent to organic fouling, as shown in Figure 3.4. Microorganisms can easily accumulate on the gel layer that has previously been deposited on the RO membrane surface. The organic compounds around their cells are known as extracellular polymeric substances (EPS) [150]. Because of the surface-active properties of EPS, EPS will remain as a matrix around the microbes, acting as a kind of blanket for microbes to sit inside, and this blanket-like layer is commonly known as a biofilm. In addition, the biofilm formed on the surface of the RO membrane surface traps colloidal particles and suspended solids in the feed water, forming a layer several micrometers thick that is highly resistant to permeate flow [149].

Figure 3.4 Bio-fouled spiral-wound RO membrane.

Washing and backwashing the surface of the RO membrane is important to reduce the burden required to compensate for the loss of productivity due to biofilm formation, but this is not a straightforward task when dealing with biocontamination. The main components of the EPS, such as TEP, are polysaccharides, which means that the EPS also has a highly adhesive characteristic. Thus, each microorganism is surrounded by its own capsule of EPS during the initial formation of biofouling [151]. However, the EPS capsules coagulate with each other very quickly to form micro-colonies, and these micro-colonies become irreversibly linked with the membrane surface where the EPS capsules have adhered [152]. At this point, the initiation of washing or backwashing can only nominally achieve biofilm removal. The difficulty in controlling biofilm formation can also be observed through pretreatment. Because biofilms are a product of microbes, pretreatment methods or biocide injections can be utilized to remove the biofilm up to a 3.log reduction level to minimize the number of microbes reaching the RO membrane. Nevertheless, because the pretreatment process cannot guarantee 100% removal, the remaining 0.1% of microorganisms can enter the system, and given sufficient time, they can multiply and form a biofilm on the entire membrane surface.

Another important feature of biofilms is their ability to move to different locations. In the final stages of biofilm development, microorganisms are separated from the microbial community and dispersed into the surrounding environment, as shown in Figure 3.5. Several other mechanisms, such as quorum sensing or environmental signals, can influence the migration, but the exact cause has not yet been identified [153]. Nevertheless, the separated individual cells or clusters settle at different locations on the RO membrane surface, repeating the biofilm growth cycle.

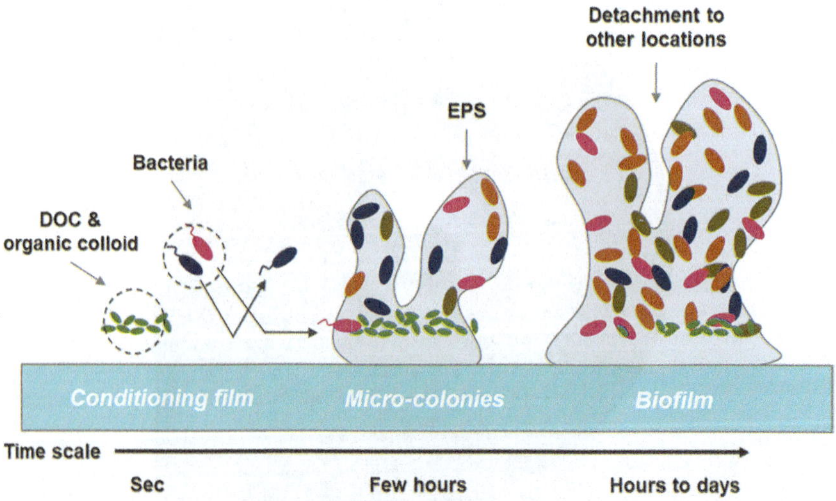

Figure 3.5 Different stages of membrane biofouling.

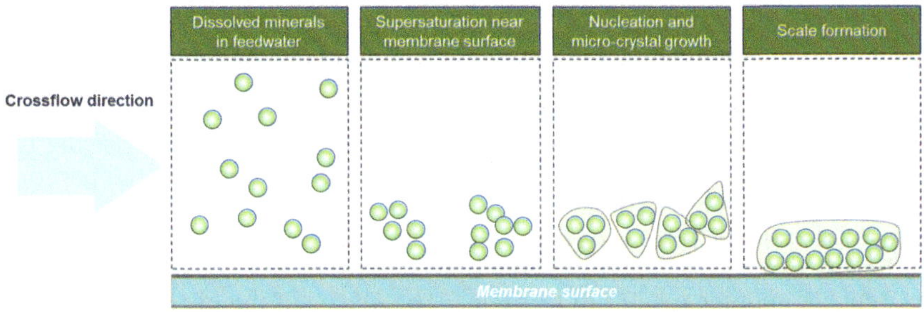

Figure 3.6 Different stages of scale formation.

3.1.4.4 Scaling

In the process of water/salt separation by the RO membrane, the salt concentration in the feed water increases significantly in the RO train. When the salt concentration exceeds the saturation limit or the solubility product of the individual salt species, it will precipitate and form a mineral scale on the membrane surface (Figure 3.6). This particle deposition on the surface of the RO membrane due to supersaturation is called scaling. Scaling is a significant problem in systems that operate under high recoveries. This trend was confirmed by the fouling phenomena observed at the surface of the RO membrane modules inside the PV. Autopsies of multiple modules confirmed that RO modules at the end of the PV suffered from more severe scaling problems than those at the front of the PV because the rear RO modules are in contact with a higher concentration than the front RO modules.

Various salts, such as calcium carbonate, calcium sulfate, barium sulfate, sodium chloride, magnesium, iron oxides, and silicate, are associated with scale formation on RO membranes. The major scalants are calcium carbonate ($CaCO_3$) and magnesium hydroxide ($Mg(OH)_2$), which are alkaline hardness scales, as well as calcium sulfate ($CaSO_4$), which is a non-alkaline hardness scale. Calcium carbonate is the most prevalent type of scalant. Initially, calcium carbonate exists as calcium and bicarbonate ions in the feed water. The formation of calcium carbonate scaling is highly influenced by the temperature and pH of the water solution, hardness of calcium, and alkalinity of bicarbonate. Because the bicarbonate in the solution is converted into carbonate under high pH conditions, which increases the possibility of calcium carbonate precipitation, the feed water pH can be adjusted to a low range of 4–6 to prevent scale formation [154]. Calcium sulfate, commonly known as gypsum, is another major cause of scaling in the RO process. Calcium sulfate has a low solubility in water under average temperature conditions of approximately 25°C and becomes more insoluble at high temperatures. As with calcium carbonate formation, the rate of gypsum precipitation increases at higher pH ranges because more sulfate species can be converted to sulfate ions [155].

Scale formation is closely related to the concentration polarization because the increased surface concentration due to the concentration polarization induces nucleation of the scaling as a result of the high degree of supersaturation. The degree of concentration polarization depends on several factors, but the most influential factors are the crossflow velocity, operating flux, and rejection. Among these factors, the crossflow velocity is very significant for controlling concentration polarization because it can be easily adjusted during RO operation. With a low crossflow velocity, more time is required for the scalants to settle on the surface of the RO membrane, thereby increasing the concentration in the immediate vicinity of the membrane. Conversely, operating the system at a high crossflow velocity exposes the RO membrane to a greater amount of scalants, increasing the probability of scalants settling on the surface of the RO membrane. Therefore, the optimal crossflow velocity is critical for the reliable maintenance of RO systems. The effects of the operating flux and rejection on the concentration polarization influence the scale formation in a similar manner. A high operating water flux results in faster concentration of the solution near the surface of the RO membrane on the feed side. A higher water flux intensifies the difference in concentration between the boundary layer at the surface of the RO membrane and the bulk feed solution. The higher the rejection of the membrane, the more solute will be retained on the surface of the RO membrane, which in turn leads to concentration of the boundary layer [156]. Concentration polarization also accelerates the formation of a cake layer or gel layer on the membrane surface. As the phenomenon progresses, problems such as solute adsorption, scaling, fouling, and pore blocking in the case of porous membrane systems ensue, resulting in the overall degradation of process performance parameters such as the water flux and salt rejection.

3.2 MEMBRANE TRANSPORT MECHANISM IN MODULE-SCALE OPERATION

In small-scale RO systems, the size of the RO membrane is very small, and the corresponding amount of water permeate is also very small. Therefore, the difference in concentration between the feed stream and outlet stream is almost negligible. In large-scale systems using large RO modules, the size of the RO membrane area is very large. Therefore, the concentration of the feed stream in the RO membrane module increases along the length of the RO module because of the large amount of water permeate. The outlet stream from the RO module contains a higher concentration than the feed stream. This means that the osmotic pressure in the RO module has a distribution, which causes an uneven water flux distribution inside the RO module. To describe these concentration and flux distributions, mass balance equations inside the RO module should be designed and merged with the RO model developed in the small-scale system. As shown in Figure 3.7, the overall mass balance equation inside the RO module can be obtained by considering the inlet flow, outlet flow, and water and salt fluxes as follows [16, 33, 157]:

$$\frac{\mathrm{d}M_{\mathrm{F}}}{\mathrm{d}t} = \frac{u_{\mathrm{F,in}}WH}{2}\rho_{\mathrm{F,in}} - \frac{u_{\mathrm{F,out}}WH}{2}\rho_{\mathrm{F,out}} - J_{\mathrm{w}}(W\Delta x)\rho_{\mathrm{w}} - J_{\mathrm{s}}(W\Delta x), \tag{3.23}$$

where M_{F} is the overall mass of the feed solution; t is the time domain; u is the linear velocity; ρ is the density; W, $H/2$, and Δx are the channel width, height, and length, respectively; and the subscripts F,in, F,out, and w denote the feed

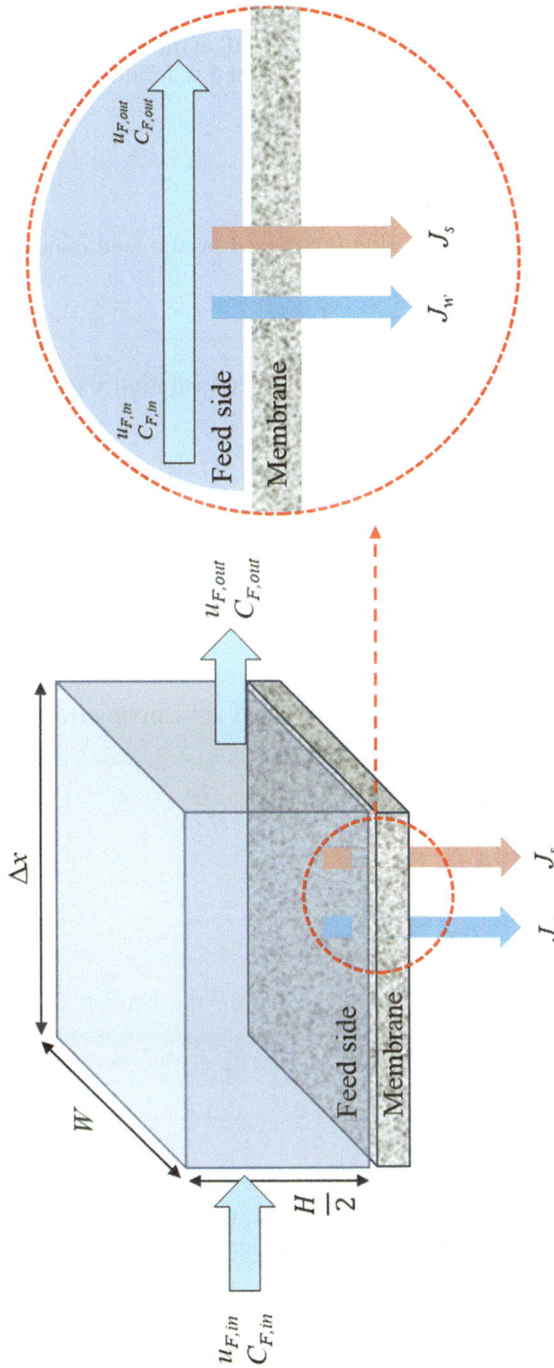

Figure 3.7 Schematic of the inlet flow, outlet flow, and flux terms in the RO module.

inlet, feed outlet, and water, respectively. Assuming a steady state, the rate of accumulation term becomes zero. By dividing all of the terms by $W \cdot H/2 \cdot \Delta x$, the differential term of the linear velocity is obtained as follows:

$$\frac{\mathrm{d}(u_F \rho_F)}{\mathrm{d}x} = -\frac{J_w \rho_w}{H/2} - \frac{J_s}{H/2}, \tag{3.24}$$

$$\rho_F \frac{\mathrm{d}u_F}{\mathrm{d}x} + u_F \frac{\mathrm{d}\rho_F}{\mathrm{d}x} = -\frac{J_w \rho_w}{H/2} - \frac{J_s}{H/2}. \tag{3.25}$$

The density of the feed solution is affected by the feed concentration. Eq. (3.25) can be rearranged as follows:

$$\rho_F \frac{\mathrm{d}u_F}{\mathrm{d}x} + u_F \frac{\mathrm{d}\rho_F}{\mathrm{d}C_F} \frac{\mathrm{d}C_F}{\mathrm{d}x} = -\frac{J_w \rho_w}{H/2} - \frac{J_s}{H/2}. \tag{3.26}$$

Similarly, the salt mass balance of the feed solution can be obtained as follows [16, 33, 157]:

$$\frac{WH\Delta x}{2} \frac{\mathrm{d}C_F}{\mathrm{d}t} = \frac{u_{F,in} WH}{2} C_{F,in} - \frac{u_{F,out} WH}{2} C_{F,out} - J_s(W\Delta x). \tag{3.27}$$

At steady state, the differential term of the feed concentration can be obtained by dividing all terms by $W \cdot H/2 \cdot \Delta x$ as follows:

$$\frac{\mathrm{d}(u_F C_F)}{\mathrm{d}x} = u_F \frac{\mathrm{d}C_F}{\mathrm{d}x} + C_F \frac{\mathrm{d}u_F}{\mathrm{d}x} = -\frac{J_s}{H/2}, \tag{3.28}$$

$$u_F \frac{\mathrm{d}C_F}{\mathrm{d}x} = -\frac{J_s}{H/2} - C_F \frac{\mathrm{d}u_F}{\mathrm{d}x}. \tag{3.29}$$

By combining Eqs. (3.26) and (3.29), Eq. (3.26) can be arranged as follows [16, 158]:

$$\rho_F \frac{\mathrm{d}u_F}{\mathrm{d}x} + \frac{\mathrm{d}\rho_F}{\mathrm{d}C_F}\left(-\frac{J_s}{H/2} - C_F \frac{\mathrm{d}u_F}{\mathrm{d}x}\right) = -\frac{J_w \rho_w}{H/2} - \frac{J_s}{H/2}, \tag{3.30}$$

$$\frac{\mathrm{d}u_F}{\mathrm{d}x} = \frac{-\frac{J_w \rho_w}{H/2} - \frac{J_s}{H/2}\left(1 - \frac{\mathrm{d}\rho_F}{\mathrm{d}C_F}\right)}{\rho_F - C_F \frac{\mathrm{d}\rho_F}{\mathrm{d}C_F}}. \tag{3.31}$$

Eqs. (3.29) and (3.31) are merged to obtain the explicit differential term about C_F.

$$\frac{\mathrm{d}C_F}{\mathrm{d}x} = -\frac{J_s}{u_F H/2} - \frac{-\frac{C_F J_w \rho_w}{u_F H/2} - \frac{C_F J_s}{u_F H/2}\left(1 - \frac{\mathrm{d}\rho_F}{\mathrm{d}C_F}\right)}{\rho_F - C_F \frac{\mathrm{d}\rho_F}{\mathrm{d}C_F}} \tag{3.32}$$

If the concentration of the solution is low, the feed solution density can be approximated as $\rho_F = \rho_w + C_F$. This assumption is very helpful for simplifying Eqs. (3.31) and (3.32) [16, 158, 159].

$$\frac{\mathrm{d}u_F}{\mathrm{d}x} = -\frac{J_w}{H/2} \tag{3.33}$$

$$\frac{dC_F}{dx} = -\frac{J_s}{u_F H / 2} + \frac{C_F J_w}{u_F H / 2} \tag{3.34}$$

In addition, the pressure of the feed solution is reduced by the friction loss in the channel. The pressure drop caused by the friction loss should be considered in estimating the water flux inside the RO module. The pressure drop inside the RO module can be obtained using the pipeline friction correlation as follows [160]:

$$\frac{dP_F}{dx} = -\frac{2\rho f u^2}{d_h}. \tag{3.35}$$

Then, the linear velocity, concentration, and pressure change of the feed solution along the RO module length are calculated using differential equations (3.33), (3.34), and (3.35). The linear velocity change inside the RO module is directly correlated to the water flux in the RO module. Thus, the permeate water production rate in the RO module and RO system can be estimated for the applied pressure. The recovery is also calculated based on the permeate water production rate and feed flow rate. The differential equation for the feed concentration can be used to estimate the salt passage across the RO membrane. The salt rejection in the RO system can then be calculated. Using the derived model equations, all of the important performance indicators for RO, such as recovery and rejection, can be obtained.

However, these model equations are derived based on a 1-D simulation. It is assumed that the velocity and concentration profiles along the membrane width are uniform. Even though this assumption is quite reasonable for various industrial RO modules such as spiral-wound types, a more detailed explanation and modeling approach are needed to estimate the velocity profile on the surface of the spacer. In this case, 2-D or 3.D simulations should be developed. Nevertheless, the 1-D simulation is a relevant modeling approach for a module-scale RO system.

3.3 ENERGY CONSUMPTION MODEL IN THE RO PROCESS

A high-pressure pump (HPP) is required to raise the pressure of the feed stream in the RO system. Electrical energy must be supplied to run the HPP. In other words, most of the energy consumption in an RO system is electricity. Thus, hereinafter, the electrical energy consumption is simply denoted as the energy consumption. The energy consumption of the HPP is usually dependent on the feed flow rate and pressure. To calculate the energy consumption of an RO system, the feed flow rate and applied pressure should be carefully estimated. From the model equations suggested in Sections 3.1 and 3.2, the applied pressure and linear velocity of the feed stream can be calculated. To satisfy the target recovery of the desired RO system, the applied pressure and linear velocity are adjusted. The linear velocity is directly correlated to the volumetric flow rate. The SEC of the RO process is calculated based on the volumetric flow rate and pressure of the feed stream.

In addition, the energy recovery device (ERD) should be considered when estimating the energy consumption of the RO system. Because the retentate stream after the RO module still contains a large amount of pressure energy [161], the pressure energy should be recovered to reduce the overall energy consumption of the RO system. Thus, the extent to which the pressure energy in the retentate can be recovered should be calculated. The energy efficiency of ERDs has been developed for systems from turbine-type ERDs such as the Pelton turbine to isobaric ERDs such as pressure exchangers [11]. Recently, the energy efficiency of pressure exchangers has exceeded 95% [162]. Because the common SWRO system is operated at a recovery of 50%, the retentate stream is almost half of the feed stream. The energy remaining in the retentate stream is quite large; therefore, a high energy efficiency of the ERD is very important. Therefore, the installation of highly efficient ERDs in RO systems is indispensable for minimizing the SEC. Because ERDs are not perfect, the pressure loss should be supplemented by an additional pump. To compensate for the pressure loss caused by the ERD efficiency and pressure drop in the RO module, a booster pump (BP) is placed after the ERD. A flow diagram of the RO process with an ERD and HPP is shown in Figure 3.8.

The RO energy consumption model can be derived from the applied pressure (P_{RO}) and feed flow rate (F_{feed}) as follows [16, 17, 33, 159]:

$$E_{RO} = \frac{F_{feed}}{\eta_{pump}}(P_{RO} - P_0) = \frac{F_p}{\eta_{pump} r_{RO}}(P_{RO} - P_0) \quad \text{(without ERD)}, \tag{3.36}$$

where η_{pump} is the pump efficiency, r_{RO} is the RO recovery, F_p is the freshwater production rate, and P_0 is the ambient pressure. This equation is for the RO system without an ERD. In the case of an RO system with an ERD, the recovered pressure energy should be calculated. The amount of energy recovered in the ERD is obtained using a pressure balance equation [159].

$$\eta_{ERD} F_p \left(\frac{1}{r_{RO}} - 1 \right)(P_{RO} - P_{fric} - P_0) = \frac{F_p}{r_{RO}}(P_{ERD} - P_0), \tag{3.37}$$

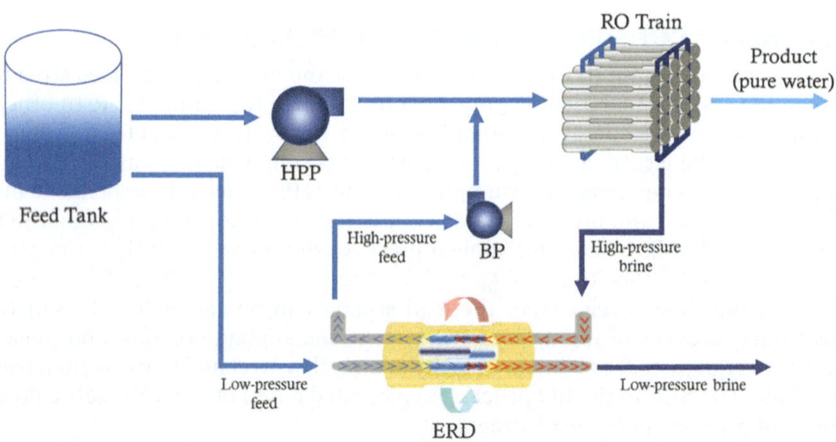

Figure 3.8 Flow diagram with a pump, ERD, and RO module in the RO process.

where η_{ERD} is the ERD efficiency, P_{fric} is the pressure drop due to friction loss in the RO module, and P_{ERD} is the increased feed pressure achieved by recovering the pressure energy in the retentate stream. The left-hand side of Eq. (3.37) is the recovered pressure energy, and the right-hand side is the increased amount of pressure energy in the feed stream. The energy consumption of the RO system with an ERD is then derived as follows:

$$\begin{aligned} E_{RO} &= \frac{F_{feed}}{\eta_{pump}}(P_{RO} - P_{ERD}) \\ &= \frac{F_p}{\eta_{pump} r_{RO}}[P_{RO} - \eta_{ERD}(1 - r_{RO})(P_{RO} - P_{fric} - P_0) - P_0] \quad \text{(with ERD).} \end{aligned} \tag{3.38}$$

Finally, the SEC of the RO process can be obtained by dividing the energy consumption of the RO system by the freshwater production rate as follows:

$$SEC_{RO} = \frac{E_{RO}}{F_p} = \frac{1}{\eta_{pump} r_{RO}}(P_{RO} - P_0) \quad \text{(without ERD),} \tag{3.39}$$

$$SEC_{RO} = \frac{E_{RO}}{F_p} = \frac{1}{\eta_{pump} r_{RO}}[P_{RO} - \eta_{ERD}(1 - r_{RO})(P_{RO} - P_{fric} -_0 P) - P_0] \quad \text{(with ERD).} \tag{3.40}$$

Note that the SEC of the RO system with an ERD is dependent on the recovery. A high-recovery RO system has a small retentate stream. For example, an RO system with 80% recovery discharges only 20% of the total amount of feed stream into the retentate stream. Thus, the pressure energy recovered from the retentate stream is not significant in the high-recovery RO system. Some installed desalination plants with high recovery (over 80%) do not use ERDs because they cannot provide any significant benefit to the desalination plant. In the SWRO system, however, the recovery is approximately 40%–50%; therefore, ERD utilization is very important.

doi: 10.2166/9781789061215_0055

Chapter 4

Recent trends in the SEC of SWRO

4.1 SWRO PLANTS WORLDWIDE

4.1.1 Data collection

Information on the design specifications and specific energy consumption (SEC) of RO plants was collected from 61 large-size SWRO desalination plants (with capacities greater than or equal to 10 000 m^3/d) commissioned after 2000 [11]. Numerous sources were utilized for the analysis, such as academic journals, books, company catalogs, and national reports. Because each reference presented similar but slightly different data, the data were cross-checked against other references if available. Academic journals were regarded as the most reliable data source, and reports authored by personnel of a specific SWRO plant were also preferred. In addition, the use of operational data was preferred over design values.

4.1.2 Data processing

Countries were classified into four regions: the Americas, Asia/Pacific, Middle East North Africa (MENA)/Mediterranean (excluding the Arabian Gulf), and the Arabian Gulf. The Arabian Gulf region is separate from the MENA/Mediterranean region, as the feeds for this region have high salinity and high temperature, different from those in other MENA/Mediterranean countries. The commissioning year was determined based on the operation year with the corresponding capacity, and the capacity was rounded to the nearest tenth.

The average values were used for feed data (e.g., TDS and temperature), but unobtainable data were assumed using the feed characteristics at the intake points or those of the nearest SWRO plants. However, the feed characteristics fluctuate in regions with significant seasonal variation, which reduces the accuracy of the analysis. The feed TDS values were rounded to the nearest hundredth, and the temperatures were rounded to the nearest whole number. The product TDS from single-pass systems was adopted without classifications, but that from two-pass systems was differentiated as before- and after-remineralization. When TDS data were presented as conductivity, they was converted to mg/L using the relationship 1000 μs/cm = 640 mg/L.

The average values rounded to the nearest whole number were used for the recovery. The recoveries of SWRO and BWRO were distinguished, but the overall recovery of the plant was expressed for SWRO if detailed data were unavailable. When the recovery value was not obtainable, values of 45% and 90% were assigned for the SWRO and BWRO processes, respectively. The types of energy recovery devices (ERDs) were classified as Pelton turbine (PT), dual work exchanger energy recovery (DWEER), and pressure exchanger (PX) in the analysis regardless of the various models from different companies. Finally, the SECs of the plant and RO system were differentiated, and the average values rounded to one decimal place were utilized.

4.1.3 Trends in datasets

The data associated with the commissioning years (Table 4.1) were utilized to investigate the trends in SWRO applications. An increasing number of SWRO applications and their capacities can indicate an increasing demand for desalination over the years. Moreover, the development of high-efficiency ERDs can be represented in the reduction in SEC over time. Finally, the trends in RO configurations can be analyzed through the required water quantity and quality in the SWRO process.

Different types of graphs were used to illustrate the relationships between factors. If auxiliary lines were provided, the trends in the data were clearly observed. Thus, the prediction interval (PI) was generated with 95% confidence, which addresses the prediction accuracy of the targets and thus the trends in the datasets [163, 164]. Further, ellipses were created for groups with 95% confidence to visually present the correlation among the data [165].

4.2 INCREASING LARGE-SCALE SWRO APPLICATIONS

SWROs provided only 10% of the global seawater desalination capacity in 1999 [281]. In this period, SWRO technologies were immature (i.e., low energy efficiency) and applied in small-size desalination plants. However, SWRO desalination has conquered the mainstream of the desalination field in the 2000s. With successful experiences in SWRO operation with low SEC, numerous large-size SWRO desalination plants have been built (Figure 4.1).

The application of SWRO has increased dramatically in the MENA/Mediterranean region. The use of desalinated water is inevitable owing to the dry weather and lack of freshwater sources in these regions, and SWRO has been widely adopted owing to its high energy efficiency compared with thermal-based methods (e.g., MSF or MED). SWRO applications are common in Spain and Israel, where SWRO plants utilize seawater from the Mediterranean Sea. Saudi Arabia also operates SWRO plants by supplying seawater from the Red Sea.

Table 4.1 Specifications of SWRO desalination plants. Different feed conditions, equipment, and RO configurations are employed in each plant, affecting the SEC of the RO system and the plant.

Plant	Region	Country	Commissioning year	Capacity (m^3/d)	Feed TDS (mg/L)	Product TDS (mg/L)	Temp. (°C)	Pre-pre-treatment	Pretreatment
Los Cabos	Americas	Mexico	2006	17 300	30 900	<400[pl]	23[nr]	–	DMF
Carlsbad	Americas	USA	2015	190 000	33 500	N/A	23	–	DMF
El Coloso	Americas	Chile	2006	45 400	36 500	400	17	DAF	DMF
La Chimba	Americas	Chile	2003	52 000	36 000	N/A	17	–	DMF
W.E.B. #2	Americas	Aruba	2012	24 000	36 000	15	28	–	–
Curacao	Americas	Curacao	2000	10 200	37 900	<40[b]	29[nr]	–	DMF
Point Lisas	Americas	Trinidad and Tobago	2002	146 000	35 000	85	29	–	DMF
Caofeidian	Asia/Pacific	China	2012	50 000	36 000	N/A	14	DAF	UF
Yuhuan Power Plant	Asia/Pacific	China	2006	34 600	34 000	<20[b,g]	24	–	UF
Fukuoka	Asia/Pacific	Japan	2005	50 000	35 000	<200[a]	20	–	UF
Gijang	Asia/Pacific	South Korea	2014	45 500	36 000	<250[a]	19	DAF	UF/DMF
Adelaide (Port Stanvac)	Asia/Pacific	Australia	2012	300 000	37 200	<200[a]	18	–	UF
Gold Coast	Asia/Pacific	Australia	2009	133 000	38 000	<220[a]	23	–	DMF
Melbourne (Victorian)	Asia/Pacific	Australia	2012	450 000	37 100	<300[a]	16	–	DMF
Perth I (Kwinana)	Asia/Pacific	Australia	2006	143 700	36 500	<200[a]	20	–	DMF
Perth II (Southern)	Asia/Pacific	Australia	2013	306 000	36 500	<200[a]	20	–	UF
Sydney	Asia/Pacific	Australia	2010	250 000	36 500	<115[a]	20	–	DMF
Nemmeli	Asia/Pacific	India	2013	100 000	40 000	300	28	–	UF
Singspring	Asia/Pacific	Singapore	2005	136 400	31 600	<250[a]	31	DAF	DMF

(Continued)

Table 4.1 Specifications of SWRO desalination plants. Different feed conditions, equipment, and RO configurations are employed in each plant, affecting the SEC of the RO system and the plant. *(Continued)*

Plant	Region	Country	Commissioning year	Capacity (m^3/d)	Feed TDS (mg/L)	Product TDS (mg/L)	Temp. (°C)	Pre-pre-treatment	Pretreatment
Tuaspring	Asia/Pacific	Singapore	2013	318 500	31 600	N/A	31	–	UF
Sadara	Arabian Gulf	Saudi Arabia	2016	149 000	<45 000	<40[b]	27	DAF	UF
Ras Abu Fontas A3	Arabian Gulf	Qatar	2017	164 000	45 900	N/A	25	DAF	UF
Fujairah	Arabian Gulf	UAE	2004	171 000	38 300	<120[b]	29	–	DMF
Ghalilah	Arabian Gulf	UAE	2015	68 000	42 000	<500	27	DAF	UF
Khorfakkan	Arabian Gulf	UAE	2008	23 000	37 400	N/A	28	–	DMF
Layyah	Arabian Gulf	UAE	2008	23 000	37 400	<450	28	DAF	DMF
Palm Jumeirah	Arabian Gulf	UAE	2005	64 000	42 000	N/A	28	–	UF
Arucas–Moya	Mediterranean	Spain (The Canary Islands)	2009	15 000	37 100	<400	25	–	DMF
Las Palmas III	Mediterranean	Spain (The Canary Islands)	2011	86 000	37 500	260	25[nr]	–	DMF
Maspalomas II	Mediterranean	Spain (The Canary Islands)	2001	26 200	39 400	<350	25[nr]	–	DMF
Dhekelia	Mediterranean	Cyprus	2004	40 000	41 800	350	25	–	DMF
Larnaca	Mediterranean	Cyprus	2001	54 000	40 000	300[b]	23	–	DMF
Aguilas–Guadalentin	Mediterranean	Spain	2013	181 000	39 500	<400	20	–	DMF
Alicante I	Mediterranean	Spain	2006	66 000	38 000	<500	21	–	DMF
Alicante II	Mediterranean	Spain	2008	65 000	38 000	<400	21	–	DMF
Almeria City	Mediterranean	Spain	2001	50 000	39 000	<400	20	–	DMF
Barcelona (Llobregat)	Mediterranean	Spain	2009	200 000	39 700	<130[b]	19	–	DMF
Campo de Dalias	Mediterranean	Spain	2015	97 200	39 800	N/A	19	–	DMF

Carboneras	Mediterranean	Spain	2005	120 000	38 000	<400	22	–	DMF
Rambla Morales	Mediterranean	Spain	2006	60 000	39 000[nr]	N/A	20[nr]	–	DMF
San Pedro del Pinatar I	Mediterranean	Spain	2005	65 000	40 900	<400	23	–	DMF
San Pedro del Pinatar II	Mediterranean	Spain	2006	65 000	40 900	N/A	23	–	DMF
Tordera	Mediterranean	Spain	2002	28 000	37 500	<500	19[nr]	–	DMF
Torrevieja	Mediterranean	Spain	2013	240 000	40 900[nr]	N/A	23[nr]	–	DMF
Valdelentisco	Mediterranean	Spain	2010	140 000	39 000	<400	21	–	DMF
Beni Saf	MENA	Algeria	2009	200 000	36 500	N/A	24	–	DMF
Honaine	MENA	Algeria	2011	200 000	35 900	N/A	21	–	DMF
Skikda	MENA	Algeria	2009	100 000	39 300	<450	23	–	DMF
Marsa Matrouh (Baghoush)	MENA	Egypt	2013	24 000	38 700	200	29[nr]	–	DMF
Jorf Lasfar	MENA	Morocco	2014	75 800	38 000	N/A	30	DAF	UF
Al Dur	MENA	Bahrain	2012	218 000	47 000	<200[a]	28	DAF	DMF
Ashkelon	MENA	Israel	2005	330 000	40 700	<80[b]	23	–	DMF
Hadera	MENA	Israel	2010	350 000	40 700	<270[a]	23	–	DMF
Palmachim 1	MENA	Israel	2007	110 000	42 000	<300[a]	25	–	DMF
Sorek (Soreq)	MENA	Israel	2013	540 000	40 800	<300[a]	27	–	DMF
Kindasa	MENA	Saudi Arabia	2006	27 000	42 500	<250[a]	30	–	UF
Rabigh IWSPP	MENA	Saudi Arabia	2009	168 000	39 600	<10[b]	29	–	DMF
Shuaibah III	MENA	Saudi Arabia	2009	150 000	44 500	45[b]	30	–	DMF
Shuqaiq II	MENA	Saudi Arabia	2009	212 000	37 100	30[b]	29	–	DMF
Yanbu	MENA	Saudi Arabia	2006	50 000	46 400	<500	28	–	DMF
Sur	MENA	Oman	2009	80 000	39 700	123[a]	28[nr]	DAF	DMF

(*Continued*)

Table 4.1 Specifications of SWRO desalination plants. Different feed conditions, equipment, and RO configurations are employed in each plant, affecting the SEC of the RO system and the plant. (*Continued*)

Plant	Overall configuration	SWRO configuration	SWRO recovery (%)	BWRO configuration	BWRO recovery (%)	ERD type	Plant SEC (kWh/m^3)	RO system SEC (kWh/m^3)	References
Los Cabos	Single pass	Single stage	49	–	–	PX	3.5	N/A	[166–168]
Carlsbad	SPSP	Single stage	50 (overall)	Cascade[c]	N/A	PX	3.6	N/A	[169–171]
El Coloso	Single pass	Single stage	50	–	–	PT	4.3[h]	3.4[d]	[172–174]
La Chimba	Single pass	Single stage	52	–	–	PT	4.2	N/A	[167, 175, 176]
W.E.B. #2	Full two pass	Single stage	43	Two stage	88	PX	3.8	N/A	[177–179]
Curacao	Two pass[uc]	Two stage	58 (overall)	Three stage	N/A	PT	4.2	2.6[d]	[180]
Point Lisas	Full two pass	Two stage	47	Two stage	90	PT	3.8	N/A	[181, 182]
Caofeidian	Single pass	Single stage	45	–	–	DWEER[we]	4.0	N/A	[167, 183, 184]
Yuhuan Power Plant	Full two pass	Single stage	45	Two stage	85	PX	<3.8	N/A	[167, 185–187]
Fukuoka	Partial two pass	Two stage	60	N/A	85	PT	5.5	N/A	[167, 188, 189]
Gijang	SPSP	Single stage	48	N/A	91	DWEER, TC	3.9	N/A	[190–192]
Adelaide (Port Stanvac)	Advanced SPSP[ad]	Single stage	49 (overall)	Two stage	N/A	PX	3.6	N/A	[193, 194]
Gold Coast	SPSP	Single stage	45	Two stage	85	DWEER	3.5	3.0[e]	[167, 195, 196]
Melbourne (Victorian)	Partial two pass	Single stage	48	Two stage	90	PX	<4.6	N/A	[197, 198]
Perth I (Kwinana)	Partial two pass	Single stage	45	Two stage	90	PX	3.6	2.4[d]	[199, 200]
Perth II (Southern)	SPSP	Single stage	45	Two stage	90	PX	3.5	2.9[e]	[195, 197, 201–203]
Sydney	Partial two pass	Single stage	45	N/A	N/A	DWEER	<3.9[i]	N/A	[195, 204, 205]
Nemmeli	Single pass	Single stage	45	–	–	PX	3.9	2.7[d,g]	[206–208]

Singspring	Partial two pass	Single stage	45	Two stage	90	DWEER	4.1	3.1[d]	[209–211]
Tuaspring	Partial two pass	Single stage	45	Two stage	90	DWEER	4.0	N/A	[167, 210, 212]
Sadara	Full/partial two pass[f]	Single stage	45	N/A	90	PX	4.4	3.1[e]	[213, 214]
Ras Abu Fontas A3	Two pass[uc]	Single stage	43 (overall)	Two stage	N/A	PX	4.5	N/A	[215]
Fujairah	Partial two pass	Single stage	43	Two stage	90	PT	4.5	3.8[e]	[216]
Ghalilah	Single pass	Single stage	40	–	–	PX	<3.0	N/A	[217, 218]
Khorfakkan	Single pass	Single stage	45[asm]	–	–	PX	4.0	N/A	[167, 219, 220]
Layyah	Single pass	Single stage	42	–	–	PX	4.0	N/A	[167, 219, 221, 222]
Palm Jumeirah	Full two pass	Single stage	35	Two stage	90	DWEER	4.1	N/A	[223]
Arucas–Moya	Single pass	Single stage	45	–	–	PX	3.4	N/A	[224, 225]
Las Palmas III	Single pass	Two stage	50	–	–	PX	4.1	2.3[d]	[226, 227]
Maspalomas II	Single pass	Two stage	60	–	–	PT	3.8	N/A	[180, 224]
Dhekelia	Single pass	Single stage	46	–	–	PX	5.3	N/A	[228, 229]
Larnaca	SPSP	Single stage	50	Two stage	78	PT	4.4	N/A	[230, 231]
Aguilas–Guadalentin	Partial two pass	Single stage	45	Two stage	90	DWEER	4.6	2.9[d]	[232–235]
Alicante I	Single pass	Single stage	45	–	–	PT	4.5	N/A	[181, 224, 236, 237]
Alicante II	Single pass	Single stage	43	–	–	PX	3.8	<2.5[d]	[181, 224, 236, 237]
Almeria City	Single pass	Single stage	45	–	–	PT	4.2	N/A	[167, 181]
Barcelona (Llobregat)	Partial two pass	Single stage	45	Two stage	85	PX	4.2	2.6[d]	[238–241]
Campo de Dalias	Partial two pass	Single stage	47	Two stage	87	DWEER	4.0	N/A	[242–244]

(Continued)

Table 4.1 Specifications of SWRO desalination plants. Different feed conditions, equipment, and RO configurations are employed in each plant, affecting the SEC of the RO system and the plant. *(Continued)*

Plant	Overall configuration	SWRO configuration	SWRO recovery (%)	BWRO configuration	BWRO recovery (%)	ERD type	Plant SEC (kWh/m³)	RO system SEC (kWh/m³)	References
Carboneras	Single pass	Single stage	45	–	–	PT	4.3	N/A	[181, 245–248]
Rambla Morales	Single pass	Two stage	58	–	–	PT	3.6	N/A	[220, 248, 249]
San Pedro del Pinatar I	Single pass	Single stage	45	–	–	PT	4.2	> 3.0	[236, 248]
San Pedro del Pinatar II	Single pass	Single stage	45	–	–	PX	3.8	3.1	[236, 248]
Tordera	Single pass	Single stage	45	–	–	PT	4.5	3.1[d]	[181, 248, 250]
Torrevieja	Two pass[uc]	Single stage	43 (overall)	Two stage	N/A	PX	4.7	3.0	[251–253]
Valdelentisco	Single pass	Two stage	50	–	–	PX	3.7	3.1	[254, 255]
Beni Saf	Single pass	Single stage	45	–	–	PX	4.0	2.3[d]	[256, 257]
Honaine	Single pass	Single stage	45	–	–	PX	4.4	N/A	[258–260]
Skikda	Single pass	Single stage	47	–	–	PX	3.6	N/A	[167, 221, 258]
Marsa Matrouh (Baghoush)	Single pass	Single stage	45[asm]	–	–	PX	3.1	2.2[d]	[261–263]
Jorf Lasfar	Partial two pass	Single stage	45	Two stage	87	PX	3.8	N/A	[223, 264]
Al Dur	Full two pass	Single stage	42	Two stage	90	PT	5.4	4.7[e]	[265–267]
Ashkelon	SPSP	Single stage	45	Cascade[c]	85, 85, 90	DWEER	<3.9	3.0[d]	[209, 268, 269]
Hadera	SPSP	Single stage	45	Cascade[c]	N/A	PX	4.0	2.7[d]	[107, 204, 209]

Palmachim 1	SPSP	Single stage	45	Cascade[c]	98 (total)	PT	3.5	2.9[d]	[209, 270–272]
Sorek (Soreq)	SPSP	Single stage	45 (overall)	Cascade[c]	N/A	DWEER	<4.0	2.7[d]	[204, 209]
Kindasa	Partial two pass	Two stage	50	Two stage	90	PT	4.6	3.5[e]	[273, 274]
Rabigh IWSPP	Full triple pass	Single stage	43	Two pass	90, 90	PT	4.8	N/A	[167, 220, 241, 275]
Shuaibah III	Full two pass	Single stage	42	N/A	90	PX	4.8	N/A	[167, 223]
Shuqaiq II	Full two pass	Single stage	40	PCP[pcp]	90	PT	4.4	N/A	[110, 276, 277]
Yanbu	Single pass	Single stage	39	–	–	PT	5.2[g]	N/A	[248, 278]
Sur	Two pass	Single stage	45[asm]	Two stage	N/A	DWEER[we]	3.6	N/A	[279, 280]

BVI: British Virgin Islands; DWEER: dual work exchanger energy recovery; ERD: energy–recovery device; PCP: permeate circulation process; PT: Pelton turbine; PX: pressure exchanger; SEC: specific energy consumption; SPSP: split partial second pass; TC: turbocharger; TDS: total dissolved solids; UAE: United Arab Emirates; USA: United States of America; WE: work exchanger.

[a]After the remineralization process. [b]Before the remineralization process. [c]Additional BWRO passes following two–stage BWRO to meet water standards. [d]SEC of SWRO–ERD. [e]SEC of SWRO–ERD and BWRO. [f]Full two–pass required during summer. [g]Design value. [h]Pumping system requires an additional 14 kWh/m^3 to transport water up the mountain. [i]Energy produced from the wind farm further reduced the SEC to 3.2 kWh/m^3. [ad]Front and rear permeate are each split and treated by second–pass ROs. [asm]Assumption of 45% SWRO recovery. [nr]Data from nearby plants. [pcp]Rear permeate from the BWRO second stage is circulated back to the BWRO feed to improve water standards. [pd]Pilot data. [uc]Unclassified two–pass. [we]A type of work exchanger.

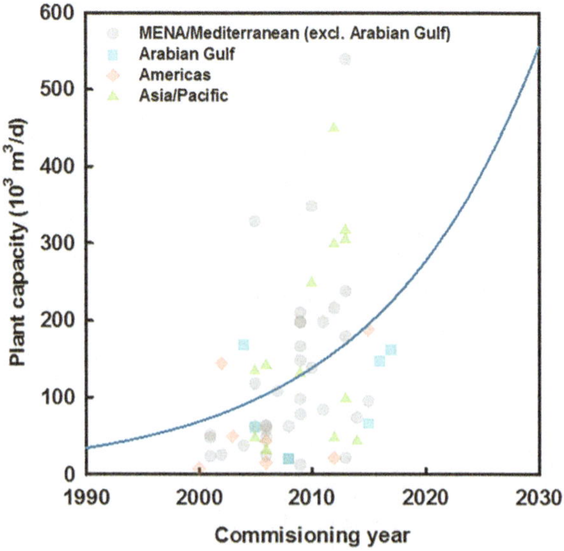

Figure 4.1 Plant capacities of SWRO desalination plants commissioned in different years. SWRO plants have been installed aggressively since 2000.

However, SWRO applications are relatively uncommon in the Arabian Gulf region, despite its largest global market share. Instead, thermal-based methods are dominant over membrane-based methods because of the high salinity of the seawater in the Arabian Gulf. Seawater in the Arabian Gulf is characterized by high salinity and high temperature, which is unfavorable for SWRO operation. In addition, heated seawater from power plants is re-utilized for freshwater production by thermal-based desalination methods. However, SWRO applications are increasing rapidly, even in the Arabian Gulf, with improvements in the efficiency of desalination equipment such as ERDs and SWRO membranes.

The market growth in the Asia/Pacific region is significant owing to increasing industrial water demand and water shortages due to climate change. In East Asia, the amount of surface water supply fluctuates owing to dramatic seasonal variations in the amount of precipitation and the temperature, and SWRO has been adopted to produce a constant supply of freshwater for industrial use. The Oceanian region also produces a large amount of water through desalination. Australia, for example, has suffered a severe drought referred to as 'Millennium Drought' [282] and installed 'the big six' desalination plants to ensure water supply. Thus, the number and capacity of desalination plants in these regions indicate the importance of desalinated water.

4.3 USING ERDS WITH HIGH ENERGY EFFICIENCY

ERDs are the key components in significantly reducing the SEC of RO systems compared with other desalting technologies. This has resulted in a dramatic

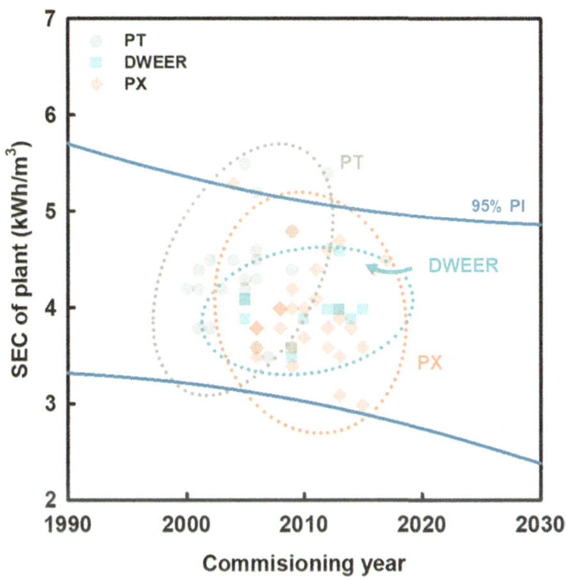

Figure 4.2 SEC of SWRO desalination plants with the improvement in ERDs.

improvement in the energy efficiency of SWRO plants over time. Because ERDs are crucial in reducing the SEC of an RO system, efforts have been made to improve the ERD efficiency (Figure 4.2). During the 2000s, PTs were developed that had improved energy efficiencies compared with Francis turbines (FTs). The PT is attached to a shaft connected to the HPP. However, turbine-type ERDs are not highly efficient owing to their dual energy conversion, in which hydraulic energy in the concentrate is converted to mechanical energy to revolve the shaft and to hydraulic energy in the feed [283, 284]. Isobaric ERDs, such as DWEERs or PXs, were thus developed to further increase the energy efficiency of the ERD. While DWEERs are applied in several plants, PXs are more dominant in applications owing to their higher efficiency [161, 285]. Recent SWRO desalination plants are assumed to adopt a PX as the ERD unless specific reasons for a different choice are provided. Figure 4.2 demonstrates how ERD development has shifted from PT to DWEER to PX over time as well as the SEC reduction achieved through this development.

4.4 INCREASING PRODUCT WATER QUANTITY

More than half of the feed is discharged as concentrate in typical low-recovery SWRO operations. Thus, efforts have been made to increase SWRO recovery by utilizing a two-stage SWRO configuration. In the 2000s, several plants adopted two-stage SWRO configurations in which the concentrate of the first-stage SWRO is fed into the second-stage SWRO. Because of the increased recovery, two-stage SWRO plants can reduce their land size as well as the flow rate of the concentrate [286].

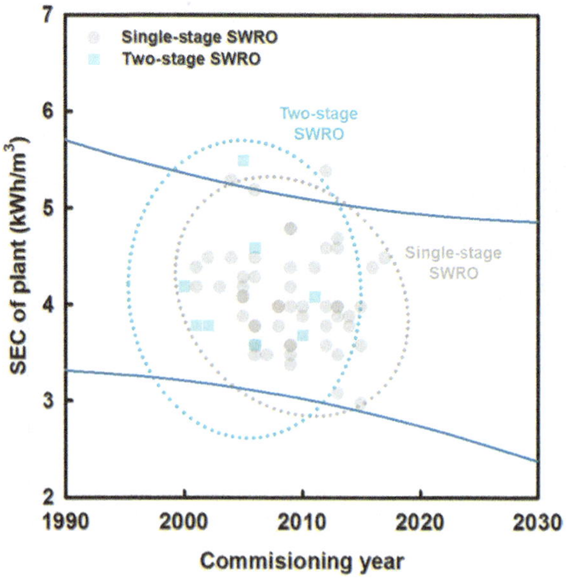

Figure 4.3 SEC of SWRO plants according to the SWRO stage configuration. Two-stage designs are adopted to increase the overall recovery.

However, two-stage SWRO configurations have higher SEC because of the extremely high-pressure operation in the second stage (Figure 4.3). In particular, a hydraulic pressure of up to 100 bar is applied in the second stage to produce freshwater from the high-salinity concentrate from the first stage [180, 286]. This also leads to high capital expenditure (CAPEX) and operating expenses (OPEX), as equipment resistant to high hydraulic pressures should be employed. Because the trend in SWRO application is toward lower SEC rather than increased recovery, two-stage SWRO configurations are not preferred in SWRO designs.

However, as the demand for zero-liquid discharge is increasing in SWRO desalination, the application of two-stage SWRO could become more frequent in the future. Additionally, high-recovery operation is more important for BWRO than SWRO, and thus BWRO systems are configured with multiple stages.

4.5 IMPROVING PRODUCT WATER QUALITY

To satisfy the demand for high-quality product, two-pass SWRO configurations have been widely adopted. SWRO plants have adopted two-pass RO, where the feed is treated by SWRO (i.e., first-pass RO) and the permeate from the SWRO is further desalinated through BWRO (i.e., second-pass RO) [17]. However, additional energy consumption is required to operate the high-pressure pumps in the additional BWRO process [17].

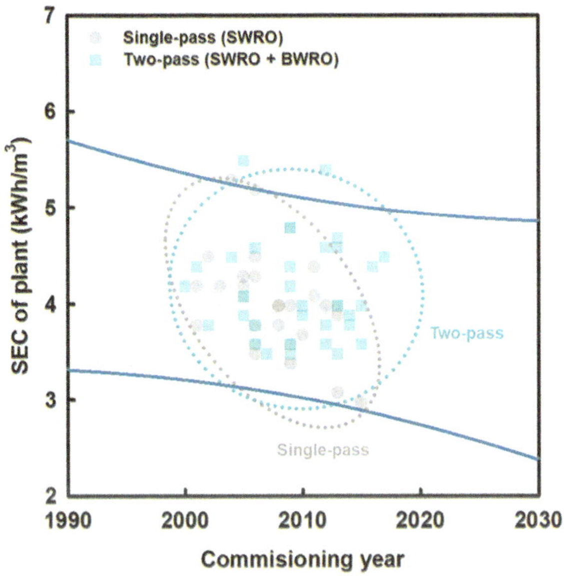

Figure 4.4 SEC of SWRO plants according to the overall RO pass configuration. Two-pass designs are used to improve the water quality.

In this regard, two-pass RO inevitably consumes more energy than single-pass RO for similar conditions. However, two-pass RO configurations exhibited lower SEC than single-pass RO configurations in the early 2000s (Figure 4.4). This is because of the types of ERDs used in those plants. During this period, the ERDs for two-pass RO were mostly isobaric types, whereas those for single-pass RO were generally turbines. When the SEC is compared for these configurations equipped with the same type of ERD, single-pass RO is more energy efficient.

Although two-pass RO is recognized for high-quality water production, triple-pass RO configurations have rarely been reported for SWRO plants for industrial water production [275]. Despite the use of multiple passes, the application of single-pass RO remains steady. Further development of SWRO membranes could improve the product quality of single-pass RO and broaden its application.

4.6 APPLYING MULTIPLE PRETREATMENT METHODS FOR HARMFUL ALGAL BLOOMS

Harmful algal blooms (HABs) can cause significant fouling problems in SWRO processes because the occurrence of HABs increases the algae and organic loads in the SWRO feed. HABs adversely affect the SWRO process by clogging the pretreatment units and deteriorating RO membranes. As a result, the permeate quality is reduced due to increased algal growth, algal organic matter (AOM) and algal toxin production, and clean-in-place and maintenance cleaning must be performed more frequently.

The incidence of HABs in the Arabian Gulf and the Gulf of Oman between 2008 and 2009 led to the closure and operational restriction of some SWRO plants in the region. Efficient and effective pretreatment methods can mitigate the occurrence of fouling resulting from HABs. This would minimize the increase in the transmembrane pressure and thus the membrane cleaning frequency.

The pretreatment method depends on the characteristics of the feed and the required permeate quality. Coagulation coupled with DMF is a conventional pretreatment method to overcome the operational challenge of HABs by removing turbidity and suspended solids [102, 287]. Owing to the ineffectiveness of DMF alone, new pretreatment methods have been introduced, such as UF plus DAF, to remove colloids and dissolved organics. DMF alone does not have the ability to satisfy the silt density index (SDI) standard for SWRO operation during HABs.

The current practice typically involves the use of DAF before DMF or pretreatment using UF. DAF is very effective for removing foulants with high buoyancy in water (such as algal cells), turbidity, color, and oils. DAF became more popular after the occurrence of HABs in the Arabian Gulf and Gulf of Oman between 2008 and 2009, and the number of DAF installations in SWRO plants has continued to grow [288]. Similarly, UF pretreatment has been reported to achieve high removal of particulate and colloidal materials that can cause fouling from the SWRO feed. UF pretreatment has also been extensively incorporated into SWRO plants worldwide [289, 290]. Compared to conventional pretreatment processes (e.g., coagulation, flocculation, sedimentation, and gravity filtration), although no significant energy savings were observed when conventional pretreatment was replaced by UF pretreatment, a significant improvement in permeate flux and recovery and reduced chemical use were reported with the UF pretreatment process [291].

Figure 4.5a shows the impact of HABs on the choice of pretreatment processes. The impact of HABs became apparent after the incidents of 2008–2009 in the Arabian Gulf and Gulf of Oman, which forced most desalination plants in the region to close, particularly those that operated a DMF process without the incorporation of a DAF unit. Although it is believed that UF pretreatment can eliminate most of the particulate matter and some of the colloidal matter, UF systems are still susceptible to fouling during HABs and the permeate from such systems would have a high fouling potential on the SWRO membrane. For instance, desalination plants that operated conventional DMF and UF during the HAB of 2008 were unable to withstand the overwhelming impact of the blooms [292]. This led to the introduction of the DAF pretreatment process as a pretreatment option during HABs owing to its proven ability to remove turbidity, oils, and greasy matter.

DAF was rarely used in SWRO plants before the challenges of HABs, but it has continued to gain attention owing to its effectiveness in algal-laden seawater. The effectiveness of DAF was further validated when SWRO plants without DAF systems were shut down while those with DAF systems continued

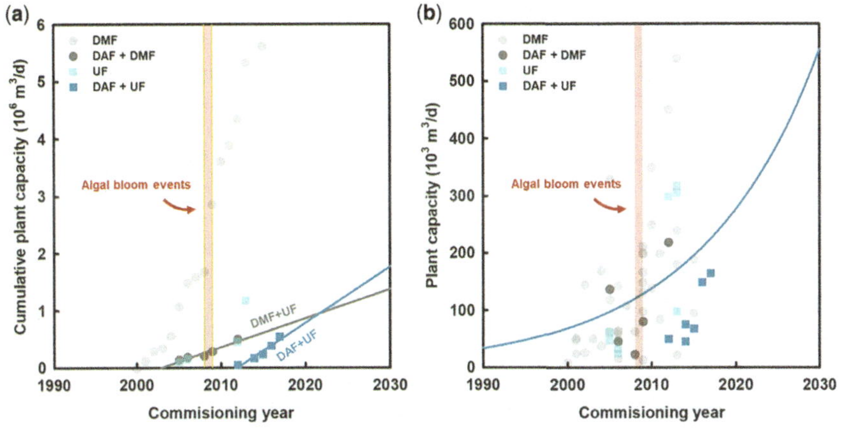

Figure 4.5 Pretreatment process modifications due to HAB occurrence: (a) single pretreatment processes and (b) cumulative pretreatment processes commissioned per year.

to operate during the blooms [293]. Thus, the number of SWRO plants that have installed multiple pretreatment processes, such as DAF in combination with DMF or UF, has continuously increased (Figure 4.5b).

DAF coupled with DMF or UF are comparable pretreatment methods for SWRO plants with a production capacity of less than 100 000 m^3/d. However, a greater percentage of SWRO plants use UF pretreatment coupled with DAF, while a lesser proportion use DMF coupled with DAF (Figure 4.6a). Interestingly,

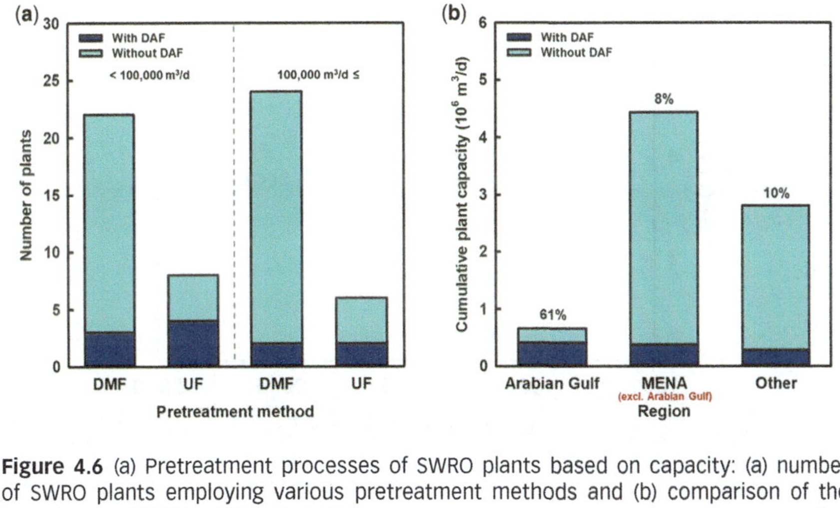

Figure 4.6 (a) Pretreatment processes of SWRO plants based on capacity: (a) number of SWRO plants employing various pretreatment methods and (b) comparison of the importance of DAF in different regions.

as the capacity of the plants increases above 100 000 m³/d, the number of plants with DMF/DAF and UF/DAF coupled systems is similar. The more frequent use of DMF in higher capacity SWRO plants is due to the robustness of the process and its cost effectiveness in terms of maintenance. The efficiency of DMF can easily be augmented by using adequate coagulation during HAB.

In the Arabian Gulf, most SWRO plants have DMF/UF pretreatment coupled with DAF (Figure 4.6b). In particular, SWRO plants subjected to UF pretreatment adopt DAF as a compulsory process. This is because the UF membrane is highly susceptible to fouling, similar to other membrane processes. Thus, a pre-pretreatment step is necessary to reduce fouling on the membrane, unlike with DMF. DAF can be a great option for UF because of its ability to reduce biofouling on the UF membrane, especially during HAB occurrence.

4.7 PERFORMING RETROFITTING AND EXPANSION

The energy efficiency of a single SWRO plant can be improved by upgrading the ERD or increasing the pump capacity (Figure 4.7). The Dhekelia SWRO Plant (Cyprus) changed its ERD from FT to PX, which contributed to a reduction in the SEC from 6.2 to 5.3 kWh/m³ [228, 229]. The Las Palmas III SWRO Plant (Canary Islands, Spain) has upgraded its facility several times. The capacity was expanded from 36 000 m³/d (1989) to 86 000 m³/d (2011), and the types of ERDs were retrofitted from FT to PX. Owing to the improvement in pump

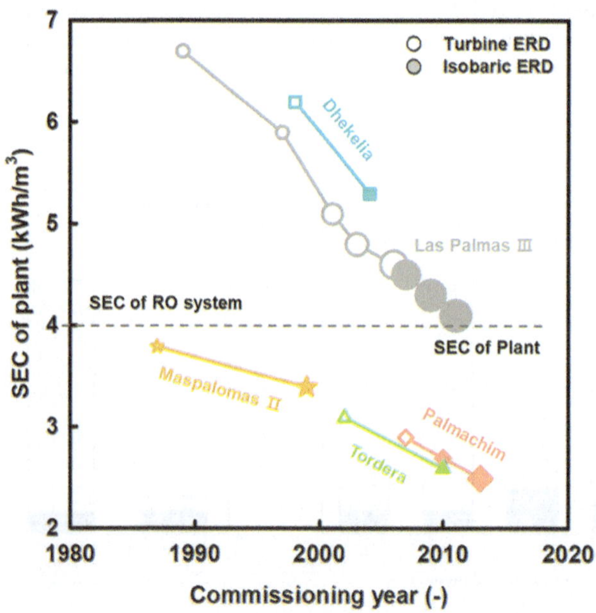

Figure 4.7 Reduction in SEC obtained by retrofitting ERDs and increasing capacity. The bubble size represents the relative capacity of the plant.

and ERD efficiencies, the SEC was reduced from 6.7 to 4.1 kWh/m^3 [226, 227]. Similarly, several studies have reported a reduction in the SEC of RO systems. Tordera SWRO Plant (Spain) reduced the SEC of its RO system from 3.1 to 2.6 kWh/m^3 by changing the ERD [181, 248, 250], and Palmachim SWRO Plant (Israel) reduced its RO SEC from 2.9 to 2.5 kWh/m^3 through an ERD retrofit and capacity expansion [209, 270–272]. The demand for low SEC values and increased water production has resulted in the retrofitting and expansion of SWRO plants.

4.8 UTILIZING RENEWABLE ENERGY

An average of 37 barrels of fossil fuel is used for SWRO desalination coupled with a power plant to desalinate 1000 m^3 of seawater. This results in the release of an average of 10 tons of CO_2 into the atmosphere [294]. Such emission levels have a significant impact on the climate, which in turn may lead to more water shortages and increased scarcity.

Future energy trends in SWRO plants are likely to focus on green and low carbon footprint desalination through the use of renewable energy. Renewable energy sources are considered the main means of addressing the greenhouse gas emissions associated with SWRO plants. Although greenhouse gases are produced during renewable energy generation, they are produced at a significantly lower level. Interestingly, a majority of SWRO desalination plants are located in the MENA region, where renewable energy sources are abundant. Thus, several SWRO desalination plants in the MENA region utilize renewable energy for water production (Figure 4.8).

Saudi Arabia aggressively utilizes photovoltaic energy to operate SWRO desalination plants. Al Khafji Plant, with a capacity of 60 000 m^3/d, is one of the largest SWRO plants that utilizes renewable energy. Because the plant is equipped with an energy storage system (ESS), photovoltaic energy can be directly utilized to operate the plant or stored in the ESS during the day. The

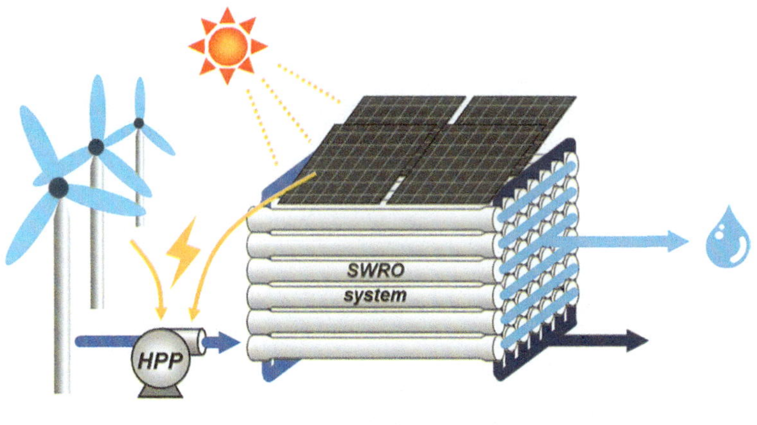

Figure 4.8 Scheme of SWRO desalination plant utilizing renewable energy.

Table 4.2. Specifications of pilot plants in the Masdar program [229].

Company	Technology	Capacity (m^3/d)
Abengoa	RO, membrane distillation (MD)	1080
Suez	RO, ion exchange (IX)	100
Sidem/Veolia	RO (using solar energy)	300
Trevi Systems	Forward osmosis (FO)	50
Mascara Renewable Water	RO (using solar energy)	30

plant is then operated with stored energy at night. In addition, a plant with a capacity of 125 000 m^3/d paired with a 4 MW photovoltaic energy system will be installed in Neom city. Likewise, a wider application of photovoltaic-energy-based SWRO plants is expected in Saudi Arabia.

Jordan has focused on the installation of renewable-energy-driven SWRO desalination plants with limited water sources. Because Jordan has a small coastline of 26 km in Aqaba, seawater desalination has not been the main source to obtain freshwater, unlike in other Middle Eastern countries. Owing to the increasing demand for freshwater, the Aqaba SWRO Plant was installed in Jordan, which produces freshwater through photovoltaic energy.

Various types of renewable energy SWRO technologies have been explored in the United Arab Emirates. In 2013, Madsar, the Abu Dhabi Future Energy Company, launched a desalination pilot program that aimed to develop desalination technologies using renewable energy. Five participating companies operated pilot plants to validate the feasibility of different desalination technologies (Table 4.2).

Renewable energy can be utilized in other regions if the feasibility of utilizing renewable energy is established. An SWRO plant in Sydney has been reported to consume less than 3.9 kWh/m^3 with a capacity of 250 000 m^3/d. However, this can be reduced further to 3.2 kWh/m^3 by utilizing the energy produced from a wind farm. Although renewable energy does not lower the amount of energy required for SWRO operation, it can decrease the energy demand from conventional energy sources.

It is necessary to supply SWRO desalination plants using renewable energy sources to achieve a water–energy nexus. SWRO paired with renewable energy facilities is expected to be a future trend in SWRO desalination. However, the feasibility of renewable energy use should be critically assessed in implementing this technology.

doi: 10.2166/9781789061215_0073

Chapter 5

Factors affecting the SEC of SWRO plants

5.1 SPECIFIC ENERGY CONSUMPTION OF PRE- AND POST-TREATMENT

Although the specific energy consumption (SEC) of pre-treatment will differ depending on the method (e.g., granular media filtration or membrane filtration), its contribution to the overall SEC of SWRO desalination plants is trivial. In contrast, the RO system accounts for the highest percentage (~71%) of energy use in an SWRO desalination plant [295]. Thus, the overall SEC of a plant is strongly affected by the SEC of the RO system.

Figure 5.1 shows the relationship between the SEC of a plant and that of the corresponding RO system based on the dataset in Table 4.1. The difference in the SEC of the plant and RO system is approximately 1 kWh/m^3 regardless of the type of ERD. In other words, the SEC for both pre- and post-treatment is nearly 1 kWh/m^3, which is not significantly affected by the RO system design. However, it should be noted that several plants require a higher SEC for intake or product pumping depending on their location. For example, several SWRO plants in Chile consume high amounts of energy for water pumping to supply desalinated water beyond the Andes Mountains.

As the RO system is the dominant energy-consuming unit, it is critical to analyze the factors associated with the SEC of the RO system. The factors affecting the SEC of the RO system also significantly influence the SEC of the plant. Feed characteristics, such as salinity and temperature, can affect the SEC of RO because they are related to the minimum energy required for operation. In addition, the SEC of RO can be determined by the target conditions, including the permeate quality and quantity. The SEC is further influenced by the efficiency of the equipment (e.g., the HPP, BP, and ERD).

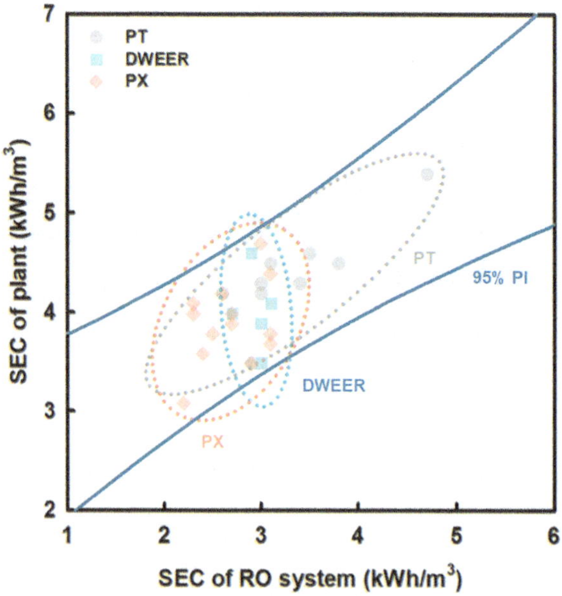

Figure 5.1 Effect of the SEC of the RO system on that of the plant. Because the RO system is the most energy-intensive unit, the SEC of the plant is strongly related to the SEC of the RO system. PT: Pelton turbine; DWEER: dual work exchanger energy recovery; PX: pressure exchanger; PI: prediction interval.

5.2 FEED CONDITIONS

5.2.1 Feed salinity

As a high-salinity feed contains high osmotic pressure, high hydraulic pressures must be applied for the production of freshwater. Thus, an increase in the feed salinity will increase the SEC of the RO system and the plant [296]. Figure 5.2 shows the effect of feed salinity on the SEC, and the trend agrees well with common sense. In fact, if the temperature effect is included, a stronger relationship would be presented for the impact of salinity on the SEC. When the feed salinity and temperature are both low, a high SEC is observed despite the low salinity. However, a high temperature will lower the SEC under high-salinity conditions (above 40,000 mg/L); thus, the SEC is underestimated. Overall, an increase in the salinity significantly increases the SEC of SWRO desalination plants.

5.2.2 Feed temperature

While the change in feed salinity influences the SEC associated with the osmotic pressure, the change in temperature affects the apparent properties of RO membranes. With an increase in the water temperature, the key parameter values of RO membranes, such as the water permeability (*A*) and salt permeability (*B*), increase [17]. At high water permeability, the SEC will

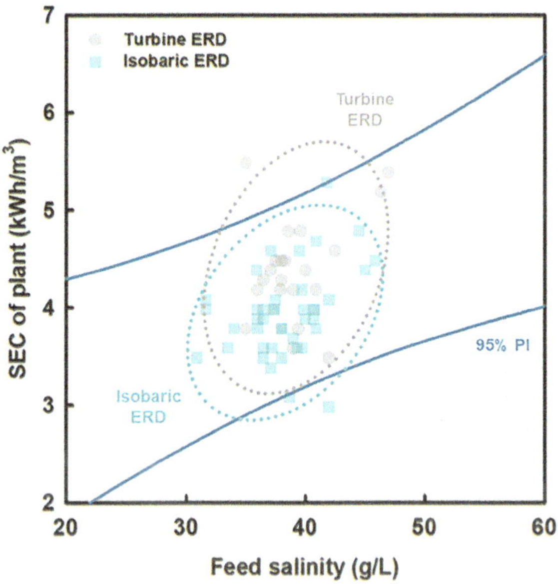

Figure 5.2 Effect of feed salinity on the SEC of SWRO desalination plants. Turbine ERD refers to PT systems, while isobaric ERD includes DWEER and PX.

be reduced because of the low hydraulic pressure operation required for water production if the other conditions are maintained. Although an increase in temperature will slightly elevate the minimum energy required for separation, its effect on increasing the SEC is relatively trivial compared to SEC reduction due to low-pressure operation [296].

Figure 5.3 does not confirm the claim that an increase in temperature can reduce the SEC of plants. A weak correlation is observed between the temperature and SEC owing to the salinity effect. However, several datasets include cases in which the feed conditions are high temperature with high salinity or low temperature with low salinity. Because the SEC is affected by the salinity and temperature, the effect of temperature on the SEC is not clearly shown in Figure 5.3. Although the effect of temperature has been investigated, it should be analyzed further by excluding the effect of salinity.

5.2.3 Overall feed conditions

Although the effects of salinity and temperature on SEC have been investigated separately, a comprehensive analysis of their effects should be performed. Thus, a mesh plot was created that presents the SEC dependence on the salinity and temperature simultaneously (Figure 5.4). As the increase in salinity and increase in temperature influence the SEC inversely, a saddle point is generated on the mesh (Figure 5.4a). This means that the SEC cannot be accurately described by analyzing the single factors alone. Thus, Figure 5.4b and 5.4c were regenerated considering the mesh plot as trends. Given that the mesh does not

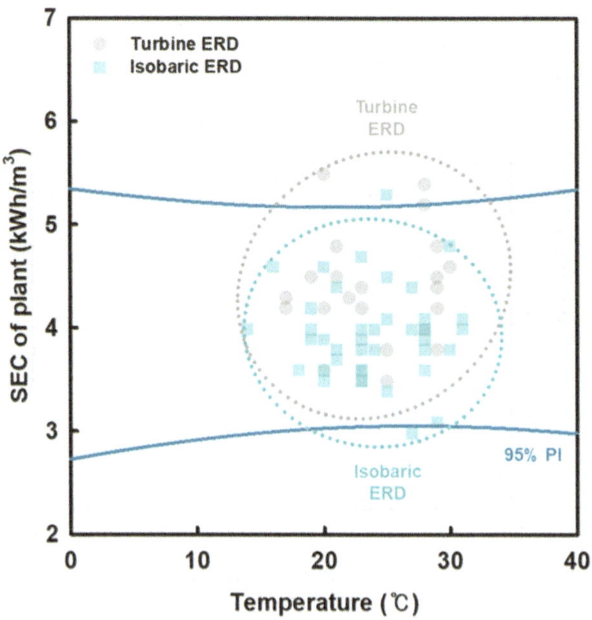

Figure 5.3 Effect of temperature on the SEC of SWRO desalination plants.

spread much in Figure 5.4b, the effect of temperature is not noticeable with the increase in salinity. In addition, the mesh in Figure 5.4c spreads significantly at high temperatures owing to the existence of various feed salinities. It is thus necessary to estimate SEC as a function of both salinity and temperature.

5.3 EQUIPMENT EFFICIENCY

5.3.1 ERD efficiency

The SEC of SWRO desalination plants, with different types of ERDs, was analyzed (Figure 5.5). SWRO plants using PTs as the ERD reported SEC values of 3.5–5.5 kWh/m^3. However, isobaric ERDs (e.g., DWEER and PX) consume less energy compared to turbine ERDs (e.g., PT). SWRO plants with DWEER reported SEC values of 3.5–4.6 kWh/m^3, while those with PXs had SEC values of 3.0–5.3 kWh/m^3. High SEC values are observed for several SWRO plants with PX systems owing to feed conditions or inferior membrane performance.

Plants with PXs accompanied by advanced technologies in membranes and other equipment operate with low SEC. Interestingly, it has a plant with a PX that has been reported to consume less than 3 kWh/m^3 for the whole plant [217, 218]. This is due to the low energy consumption of the RO system, which consists of single-pass RO accommodating high-performance membranes and employs PX as the ERD. Similarly, the SEC of the RO system greatly influences the SEC of the plant, and thus the SEC can be lowered with the development of ERD technologies.

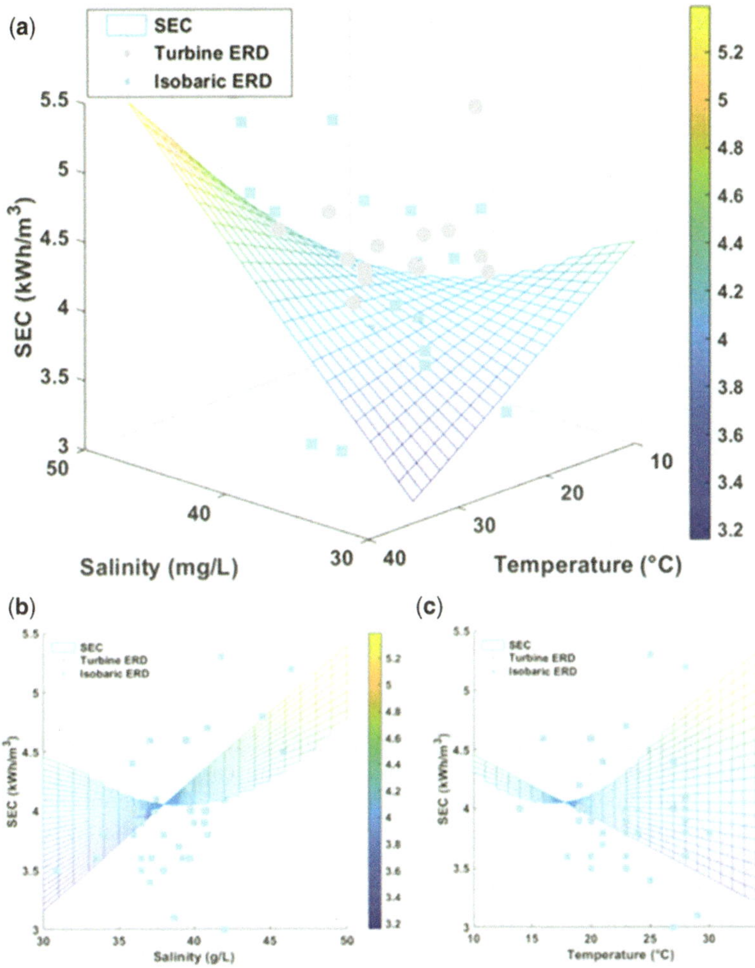

Figure 5.4 Effects of feed conditions on the SEC: (a) overall feed conditions, (b) salinity, and (c) temperature.

5.3.2 Pump efficiency

The SEC of SWRO plants can also be influenced by the efficiency of the HPP and BP. The size of the pump has been reported to be positively correlated with the pump efficiency [102]. As a result, larger-sized SWRO plants can reduce the RO SEC through improved pump efficiency. The correlation between the plant capacity and SEC is shown in Figure 5.6. Based on common sense, the SEC should decrease with increasing capacity.

However, high-capacity SWRO plants do not always exhibit a low SEC. SWRO desalination plants consist of several pump stations. The configuration

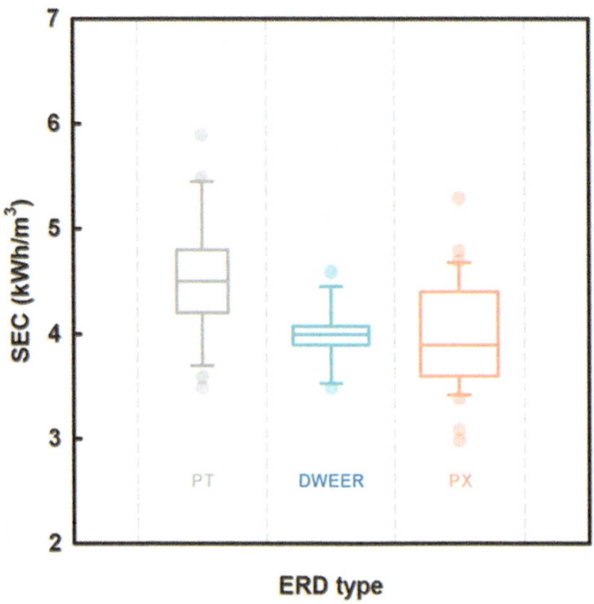

Figure 5.5 SEC of SWRO plants depending on the type of ERD. Isobaric ERDs exhibit better energy efficiency compared to turbine ERDs.

of multiple RO racks is advantageous for maintenance or emergency operations. Because the size of the pump is determined by the capacity of the pump station, the plant capacity is not strongly related to the SEC. In contrast, the relationship is apparent for SWRO plants with capacities of less than or equal to 100,000 m^3/d. This is because fewer pump stations are installed in these plants, and thus the effect is lessened.

It is advantageous to increase the size of the pump station to decrease the SEC of SWRO desalination plants. A new RO configuration has been proposed to increase the size of the pump station by integrating multiple pump stations. This configuration is referred to as the pressure center design, and the equipment of the RO system (e.g., the HPP, BP, and ERD) is placed in the middle of the train to increase its size. A large amount of feed is supplied by a larger pump, which can be more energy efficient than a typical design [295]. IDE Technologies Ltd. developed this design, and several large-size SWRO plants (Ashkelon, Hadera, and Sorek) have adopted it [107, 204, 209, 268, 269].

5.4 TARGET CONDITIONS

5.4.1 Permeate quality

The permeate quality can be improved by applying a two-pass RO design, where the feed is treated by both SWRO and BWRO. Depending on the pipe connections between the SWRO and BWRO, two-pass RO is classified as full,

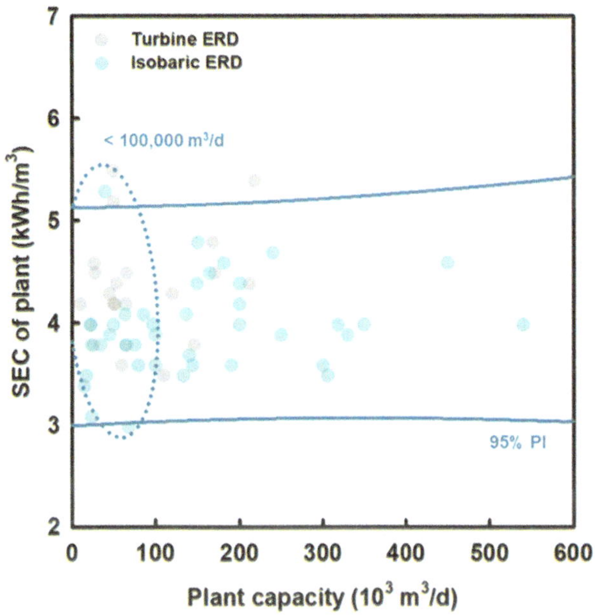

Figure 5.6 Effect of plant capacity on the SEC of SWRO desalination plants. It should be noted that the size of the pump station determines the pump efficiency.

partial, or split partial two-pass (or second-pass) RO. However, two-pass RO exhibits a higher SEC because it produces high-quality permeate. This can be explained by the higher minimum energy required for separation to produce purer water. Single-pass RO produces a permeate quality of 300–500 mg/L with a relatively low SEC (Figure 5.7). On the contrary, two-pass RO produces permeate with improved quality and higher SEC. The permeate quality from two-pass RO is 15–130 mg/L before remineralization and 115–300 mg/L after remineralization. In particular, the SEC is significantly higher when full two-pass RO is utilized to produce permeate with TDS of less than 50 mg/L (before remineralization).

The RO pass configuration will determine the SEC and permeate quality. The SEC values for different RO pass configurations are illustrated in Figure 5.8. Single-pass RO consumes relatively less energy than two-pass RO. However, several single-pass RO plants exhibit higher SEC values than two-pass RO plants, particularly when the single-pass RO plants started operation in the early 2000s. This is because low-efficiency desalination technologies were installed in these SWRO plants, including turbine ERDs and low-performance membranes. Furthermore, the SEC for two-pass RO varies based on the partial ratio of the second-pass RO. Split partial second-pass (SPSP) RO consumes less energy than other two-pass RO systems because the split design generates less mixing entropy. SPSP RO treats the SWRO rear permeate only and mixes it

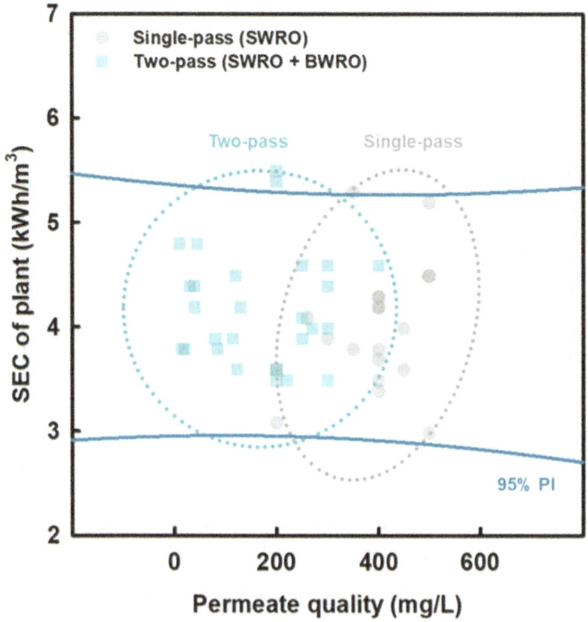

Figure 5.7 Effect of permeate quality on the SEC of SWRO plants. Two-pass RO consumes more energy to produce purer permeate.

with the SWRO front permeate to produce the final product [17]. In contrast, full two-pass RO exhibits the highest SEC among two-pass RO systems because of the larger volume treated by BWRO.

5.4.2 Permeate quantity

Recovery refers to the ratio of the permeate flow rate to the feed flow rate. A high-recovery operation is beneficial for minimizing the amount of feed required to produce a certain amount of permeate, and thus it can reduce the overall intake and pre-treatment volumes [297]. As a result, a high-recovery operation can reduce the costs of intake and brine disposal. However, the range of SWRO recovery is limited by the salinity of the feed, which is associated with the osmotic pressure. During the RO process, the osmotic pressure is constantly increasing, and the osmotic pressure reaches the hydraulic pressure through permeate production [17]. Permeate is no longer produced when the osmotic pressure reaches the hydraulic pressure [29, 33, 108]. Because SWRO recovery is limited by high osmotic pressure, SWRO plants cannot achieve high recovery with a high-salinity feed (Figure 5.9). The hydraulic pressure should be increased to achieve higher recovery; thus, several SWRO plants apply two-stage SWRO configurations.

Two-stage SWRO plants exhibit higher SEC than single-stage SWRO owing to the high-recovery operation in two-stage SWRO. The SEC of SWRO plants

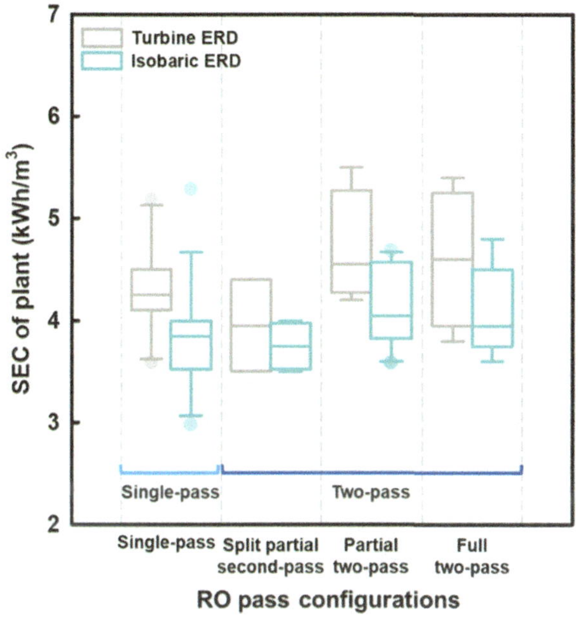

Figure 5.8 SEC according to SWRO configuration. Unclassified two-pass and triple-pass are considered as partial and full two-pass, respectively.

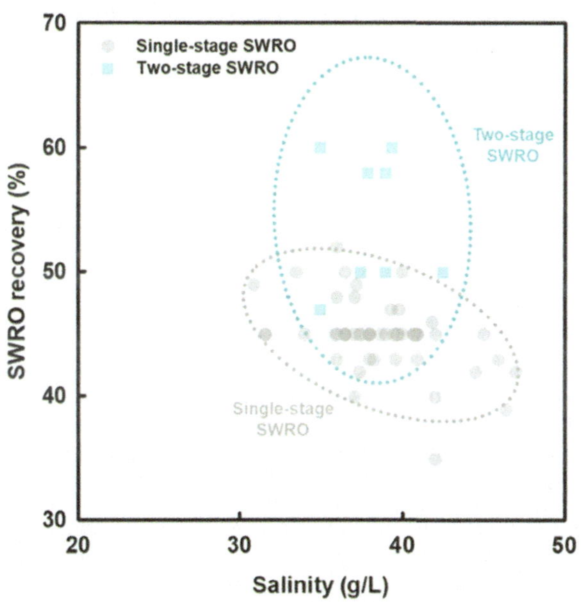

Figure 5.9 Effect of the feed salinity on SWRO recovery.

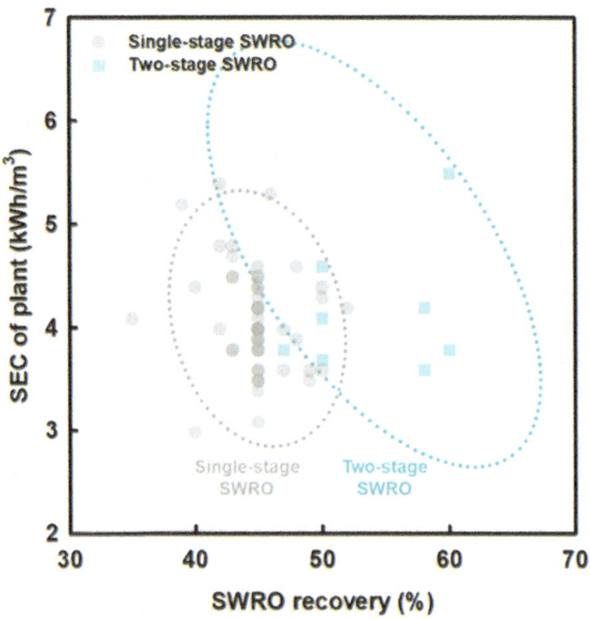

Figure 5.10 Effect of SWRO recovery on the SEC of SWRO desalination plants.

is depicted with their corresponding SWRO recovery in Figure 5.10. When the recovery of single-stage SWRO plants ranges between 40% and 50%, the SEC of these plants is relatively low. This can be attributed to an optimal recovery for the SWRO operation, which minimizes SEC [108]. In contrast, the optimal recovery for two-stage SWRO plants is approximately 60%. This is because the optimal recovery for the first stage is 40%, and an additional 20% recovery is obtained from the second stage. Interestingly, when SWRO plants are operated with a recovery less than or greater than the optimal value, the plants exhibit relatively high SEC. Therefore, it is critical to operate SWRO plants with the optimal recovery to minimize the SEC and costs.

5.5 SUMMARY AND FUTURE DIRECTIONS

With the crisis of water and energy, the SEC is considered a critical factor in the operation of SWRO desalination plants. Thus, factors associated with the SEC are systemically analyzed to understand the energy use in SWRO desalination plants. Based on this analysis, future directions are suggested to lower the SEC of SWRO plants.

- ***Dilute feed salinity***: Feed conditions such as salinity and temperature are associated with the thermodynamic minimum energy for separation. Thus, SWRO desalination plants exhibit different SEC values even when the same equipment and design are adopted. The SEC increases with an increase in feed salinity but decreases with an increase in temperature. In other words, the SEC is reduced when the salinity is lower and the temperature is higher. However, feed conditions are specific and site-dependent, and higher energy consumption may be required to control these factors. In contrast, recent dilution methods, such as forward osmosis (FO), can reduce the feed salinity with low energy by diluting seawater using water reclaimed from wastewater. Such a dilution method could lower the SEC of SWRO desalination plants if the energy required to operate FO is smaller than the energy saved in SWRO. Thus, the feasibility of dilution technologies should be investigated by considering the net amount of energy reduction.

- ***Reduce irreversible work***: The SEC of SWRO plants has been reduced significantly with the development of highly efficient equipment. The utilization of ERDs allows SWRO desalination plants to reach an SEC of 3 kWh/m^3. Currently, PX systems are the most energy-efficient ERD, with over 95% mechanical efficiency. The development of advanced ERDs would certainly lower the SEC, but the impact may not be significant. In addition, the pump efficiency can be improved when the pump size is larger. Consequently, a pressure center design has been developed to increase the pump size. However, it would not be a reasonable suggestion to reduce the SEC by increasing the pump size owing to the need for several pump stations, difficulties in maintenance, and other operational issues. Nevertheless, the SEC of the RO system can be reduced by decreasing the irreversible work in high-pressure pumps. Although both reversible and irreversible works are required by the pumps, the irreversible work is not involved in the desalination process. Therefore, reducing such wasted work can lower the SEC of RO systems and consequently that of plants.
- ***Harvest osmotic energy***: SWRO desalination plants should produce water that satisfies the required quality and quantity, and more energy is consumed to meet stricter requirements. Two-pass RO is utilized to produce high-quality permeate, and different types of two-pass RO configurations are employed depending on its purpose. Considering various RO pass configurations, the development of new RO configurations may not significantly improve the energy efficiency. On the contrary, two-stage SWRO is used to increase the overall recovery of the RO system, and it is important to operate SWRO at the optimal recovery to reduce the SEC. In summary, the reduction in SEC is focused on permeate production in terms of quality and quantity. However, it should be noted that the concentrate contains an abundance of osmotic energy, and this energy can be utilized to reduce the SEC of SWRO desalination plants if harvested. Pressured retarded osmosis (PRO) and reverse electrodialysis (RED) are promising technologies for harvesting osmotic energy from the concentrate, but their feasibility should be investigated further in combination with SWRO desalination plants.

Considering that the SEC of SWRO desalination plants is higher than the thermodynamic minimum energy, it should be possible to further reduce the

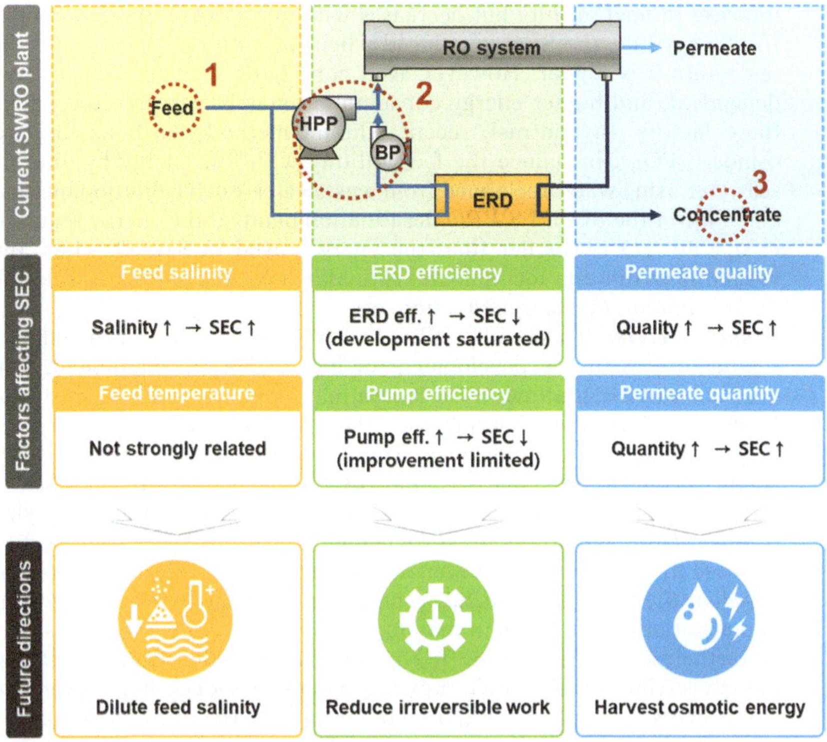

Figure 5.11 Factors affecting SEC and suggested directions for future research. (Adapted from Kim *et al.*, *Appl. Energy*, 254 (2019) 113652.)

SEC. It is expected that further SEC reduction can be achieved by improving current technologies or developing novel technologies. These technologies may include lower feed salinity, reduced irreversibility of pumps, and recovery of osmotic energy from the concentrate (Figure 5.11).

doi: 10.2166/9781789061215_0085

Chapter 6

Advanced technologies for a low-energy SWRO process

6.1 THEORETICAL ENERGY CALCULATION OF THE SWRO PROCESS

6.1.1 Exergy analysis for theoretical minimum energy

The energy consumption for the separation of seawater to obtain fresh water changes depending on the separation methodology and process equipment. For example, thermal-based desalination usually requires high specific energy consumption compared to membrane-based desalination. However, regardless of the separation method, the minimum hurdle for the separation, which is usually called the theoretical minimum energy, can be clearly identified, because it is only affected by the information of the feed and product streams. The energy efficiency of any desalination process can be calculated from the actual energy consumption to the theoretical minimum energy ratio. Therefore, the theoretical minimum energy is a criterion for evaluating the energy efficiency of the desalination process. The methodology for calculating the theoretical minimum energy is the Gibbs free energy equation for mixing [21, 45, 298], and an exergy analysis of feed and product streams [46, 299–302]. The detailed equations for calculating the theoretical minimum energy using these two methods are as follows:

$$\text{Method 1: } E_{\min} = \int -\mathrm{d}\left(\Delta G_{\mathrm{mix}}\right) = \int -RT \ln a_{\mathrm{w}} \mathrm{d}n_{\mathrm{w}} = \int \Pi_{\mathrm{s}} \bar{v}_{\mathrm{w}} \mathrm{d}n_{\mathrm{w}} \tag{6.1}$$

$$\text{Method 2: } E_{\min} = \sum \dot{X}_{\mathrm{f,out}} - \sum \dot{X}_{\mathrm{f,in}} \tag{6.2}$$

$$\dot{X}_{\mathrm{f}} = \left(h - h^{*}\right) - T_0\left(s - s^{*}\right) + \sum_{i=1}^{n} w_i \left(\mu_i^{*} - \mu_i^{0}\right) \tag{6.3}$$

where G_{mix} denotes the Gibbs function for mixing, R is the gas constant, T is the temperature, a_w is the activity of water, n_w is the number of moles of water, Π_s is the osmotic pressure of seawater, $\bar{V}_w$ is the molar volume of water, $\dot{X}_{f,out}$ and $\dot{X}_{f,in}$ are the flow exergy of the outlet and inlet streams, respectively, h is the specific enthalpy, s is the specific entropy, μ is the chemical potential, w is the mass fraction, and superscripts * and ^{0} denote the restricted dead state and the global dead state, respectively.

6.1.1.1 Method 1

In method 1, the theoretical minimum energy is obtained from the osmotic pressure of the feed solution. The correlation between the osmotic pressure and theoretical minimum energy is derived from the Gibbs free energy of mixing [21, 298]. An illustrative case is suggested to derive the theoretical minimum energy required for separation (Figure 6.1). The pure solvent and solution are placed in a solution chamber with a semi-permeable membrane in between. Owing to the existence of the osmotic pressure difference, higher pressure, which is the same as the osmotic pressure difference, should be supplied on the solution side to avoid net water flux across the membrane. At the equilibrium state, it is considered that pressure P on the pure solvent and pressure $P+\pi$ on the solution are applied. Based on solution thermodynamics, the chemical potentials in the pure solvent and solution at the equilibrium state can be expressed as follows [303, 304]:

$$\mu_A^*(P) = \mu_A(x_A, P+\pi) \tag{6.4}$$

where μ_A is the chemical potential of material A, x is the mole fraction, and the superscript * denotes pure species. To express the chemical potential of the

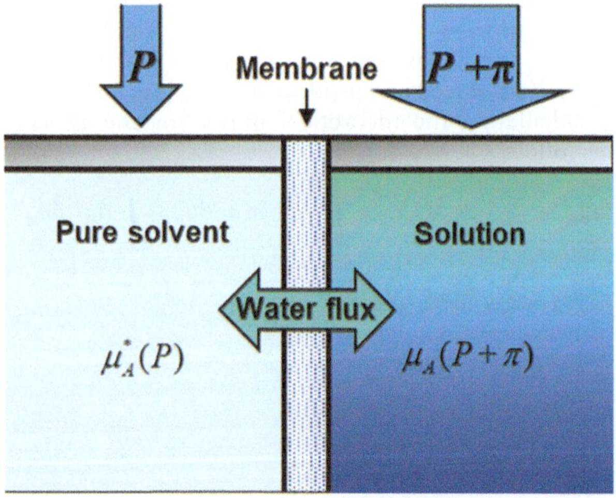

Figure 6.1 Chemical potential at the equilibrium state of the pure solvent and solution located on both sides of the RO membrane.

mixture, the chemical potential with pure species and activity coefficient are used as follows [303, 304]:

$$\mu_A(x_A, P+\pi) = \mu_A^*(P+\pi) + RT\ln(\gamma_A x_A) \tag{6.5}$$

where R is the gas constant, T is the absolute temperature, and γ is the activity coefficient. As the chemical potential is a function of pressure, it is necessary to express the pressure effect on the chemical potential. At a constant temperature, the pressure effect on the chemical potential of pure species is simplified by the integration of the volume on the pressure difference as follows [303, 304]:

$$\mu_A^*(P+\pi) = \mu_A^*(P) + \int_P^{P+\pi} V(p^*)\mathrm{d}p^* \quad \text{(at constant } T\text{)} \tag{6.6}$$

Equation (6.4) can be arranged by substituting Equations (6.5) and (6.6) as follows:

$$\mu_A^*(P) + \int_P^{P+\pi} V(p^*)\mathrm{d}p^* + RT\ln(\gamma_A x_A) = \mu_A^*(P) \tag{6.7}$$

$$-RT\ln(\gamma_A x_A) = \int_P^{P+\pi} V(p^*)\mathrm{d}p^* \tag{6.8}$$

Based on the initial state of the solutions before mixing and the final state after mixing, the Gibbs free energy of mixing can be calculated as follows:

$$\Delta G_{\mathrm{mix}} = G_{\mathrm{final}} - G_{\mathrm{initial}} \tag{6.9}$$

It is assumed that there are only two chemical species (A and B) in the membrane separation system. Initially, two pure chemical species are isolated in different chambers. After mixing these two chemical species completely, the Gibbs free energy of the mixture at the final state usually increases owing to the entropy generation based on the second law of thermodynamics. This can be attributed to changes in the chemical potential. If the pressure and temperature are consistent during mixing, the Gibbs free energy at the initial and final states can be derived from the chemical potential as follows:

$$G_{\mathrm{initial}} = n_A\mu_A^* + n_B\mu_B^* \tag{6.10}$$

$$G_{\mathrm{final}} = n_A\left[\mu_A^* + RT\ln(\gamma_A x_A)\right] + n_B\left[\mu_B^* + RT\ln(\gamma_B x_B)\right] \tag{6.11}$$

where n is the mole of each chemical. Therefore, the Gibbs free energy of mixing is derived as

$$\Delta G_{\mathrm{mix}} = n_A RT\ln(\gamma_A x_A) + n_B RT\ln(\gamma_B x_B) \tag{6.12}$$

The theoretical minimum energy for separation has the same absolute value but an opposite sign from the Gibbs free energy of mixing. As the solution system described in this chapter consists of water and salt, the differential term of the theoretical minimum energy for separation can be expressed as

$$\mathrm{d}(-\Delta G_{\mathrm{mix}}) = \mathrm{d}\left\{-RT\left[n_{\mathrm{w}}\ln a_{\mathrm{w}} + n_{\mathrm{s}}\ln a_{\mathrm{s}}\right]\right\} \tag{6.13}$$

where a is the activity, and the subscripts w and s denote water and salt, respectively. In high-performance membrane separation systems, the salt permeation is almost zero. Thus, it can be assumed that the differential term of salt permeation becomes zero for the sake of simplicity. Then, Equation (6.13) can be arranged as follows:

$$\mathrm{d}\left(-\Delta G_{\mathrm{mix}}\right) = -RT\mathrm{d}\left(n_{\mathrm{w}} \ln a_{\mathrm{w}}\right) = -RT\left(n_{\mathrm{w}}\mathrm{d}\ln a_{\mathrm{w}} + \ln a_{\mathrm{w}}\mathrm{d}n_{\mathrm{w}}\right) \tag{6.14}$$

As the change in the number of water molecules is much larger than the change in the activity of water, the differential term of water activity is negligible. Equation (6.14) can be simplified as

$$\mathrm{d}\left(-\Delta G_{\mathrm{mix}}\right) = -RT\ln a_{\mathrm{w}}\mathrm{d}n_{\mathrm{w}} \tag{6.15}$$

By incorporating Equations (6.8) and (6.15), the correlation between the theoretical minimum energy for separation and the osmotic pressure of the feed solution is obtained as follows:

$$\mathrm{d}\left(-\Delta G_{\mathrm{mix}}\right) = \pi V_{\mathrm{m}}\mathrm{d}n_{\mathrm{w}} \tag{6.16}$$

6.1.1.2 Method 2

Method 2 utilizes exergy flow calculation. The flow exergy can be calculated by the enthalpy, entropy, and chemical potential, which are functions of temperature, pressure, and concentration. The equations for calculating flow exergy can be expressed as follows [301, 302]:

$$E_{\mathrm{min}} = \sum \dot{X}_{\mathrm{f,out}} - \sum \dot{X}_{\mathrm{f,in}} \tag{6.17}$$

$$\dot{X}_{\mathrm{f}} = \left(h - h^{*}\right) - T_0\left(s - s^{*}\right) + \sum_{i=1}^{n} w_i\left(\mu_i^{*} - \mu_i^{0}\right) \tag{6.18}$$

where $\dot{X}_{\mathrm{f,out}}$ and $\dot{X}_{\mathrm{f,in}}$ are the flow exergy of the outlet and inlet streams, respectively, h is the specific enthalpy, s is the specific entropy, μ is the chemical potential, w is the mass fraction, and superscripts * and 0 denote the restricted dead state and global dead state, respectively.

The specific enthalpy of seawater is calculated using the following equations [301, 302]:

$$h\left(T, P, w_{\mathrm{s}}\right) = h\left(T, P_0, w_{\mathrm{s}}\right) + \upsilon\left(P - P_0\right) \tag{6.19}$$

$$h = h_{\mathrm{w}} - w_{\mathrm{s}}\begin{pmatrix} a_1 + a_2 w_{\mathrm{s}} + a_3 w_{\mathrm{s}}^2 + a_4 w_{\mathrm{s}}^3 + a_5 T + a_6 T^2 \\ + a_7 T^3 + a_8 w_{\mathrm{s}} T + a_9 w_{\mathrm{s}}^2 T + a_{10} w_{\mathrm{s}} T^2 \end{pmatrix} \tag{6.20}$$

$$h_{\mathrm{w}} = 141.355 + 4202.07 \times T - 0.535 \times T^2 + 0.004 \times T^3 \tag{6.21}$$

where T is the temperature (°C), P is the pressure (kPa), w_{s} is the mass fraction of salt (kg salt/kg seawater), υ is the specific volume of seawater (m^3/kg), and h_{w} is the specific enthalpy of pure water (J/kg).

The specific volume of seawater is calculated using the following equations [301, 302]:

$$v = \frac{1}{\rho} = \frac{1}{\rho_w + w_s\left(b_1 + b_2T + b_3T^2 + b_4T^3 + b_5w_sT^2\right)} \tag{6.22}$$

$$\begin{aligned}\rho_w &= 9.999\times10^2 + 2.034\times10^{-2}T - 6.162\times10^{-3}T^2 \\ &\quad +2.261\times10^{-5}T^3 - 4.657\times10^{-8}T^4\end{aligned} \tag{6.23}$$

where ρ is the density of seawater (kg/m^3) and ρ_w is the density of pure water (kg/m^3).

The specific entropy is calculated by the following equations [301, 302]:

$$s = s_w - w_s\begin{pmatrix} c_1 + c_2w_s + c_3w_s^2 + c_4w_s^3 + c_5T + c_6T^2 \\ +c_7T^3 + c_8w_sT + c_9w_s^2T + c_{10}w_sT^2 \end{pmatrix} \tag{6.24}$$

$$\begin{aligned}s_w &= 0.1543 + 15.383T - 2.996\times10^{-2}T^2 \\ &\quad +8.193\times10^{-5}T^3 - 1.370\times10^{-7}T^4\end{aligned} \tag{6.25}$$

where s_w is the specific entropy of pure water (J/kg K).

All the parameters in Equations (6.20), (6.22), and (6.24) are listed in Table 6.1.

In the case of seawater desalination, the chemical potential of water and salts is obtained by Gibbs free energy as follows [301, 302]:

$$\mu_w = \frac{\partial G}{\partial m_w} = g - w_s\frac{\partial g}{\partial w_s} \tag{6.26}$$

Table 6.1 Parameters for flow exergy calculation.

	Parameter	Value	Parameter	Value
Equation (6.20)	a_1	-2.348×10^4	a_2	3.152×10^5
	a_3	2.803×10^6	a_4	-1.446×10^7
	a_5	7.826×10^3	a_6	-4.417×10^1
	a_7	2.139×10^{-1}	a_8	-1.991×10^4
	a_9	2.778×10^4	a_{10}	9.728×10^1
Equation (6.22)	b_1	8.020×10^2	b_2	-2.001
	b_3	1.677×10^{-2}	b_4	-3.060×10^{-5}
	b_5	-1.613×10^{-5}		
Equation (6.24)	c_1	-4.231×10^2	c_2	1.463×10^4
	c_3	-9.880×10^4	c_4	3.095×10^5
	c_5	2.562×10^1	c_6	-1.443×10^{-1}
	c_7	5.879×10^{-4}	c_8	-6.111×10^1
	c_9	8.041×10^1	c_{10}	3.035×10^{-1}

$$\mu_s = \frac{\partial G}{\partial m_s} = g + (1 - w_s)\frac{\partial g}{\partial w_s} \tag{6.27}$$

where μ_w is the chemical potential of water in seawater, μ_s is the chemical potential of salts in seawater, G is the total Gibbs free energy, and g is the specific Gibbs free energy of seawater. From the definition of Gibbs free energy, the specific Gibbs free energy is calculated by the following equation:

$$g = h - (T + 273.15)s \tag{6.28}$$

Then, the differentiation of the specific Gibbs free energy is expressed as

$$\frac{\partial g}{\partial w_s} = \frac{\partial h}{\partial w_s} - (T + 273.15)\frac{\partial s}{\partial w_s} \tag{6.29}$$

$$\begin{aligned}\frac{\partial h}{\partial w_s} &= -a_1 - 2a_2 w_s - 3a_3 w_s^2 - 4a_4 w_s^3 - a_5 T - a_6 T^2 \\ &\quad - a_7 T^3 - 2a_8 w_s T - 3a_9 w_s^2 T - 2a_{10} w_s T^2\end{aligned} \tag{6.30}$$

$$\begin{aligned}\frac{\partial s}{\partial w_s} &= -c_1 - 2c_2 w_s - 3c_3 w_s^2 - 4c_4 w_s^3 - c_5 T - c_6 T^2 \\ &\quad - c_7 T^3 - 2c_8 w_s T - 3c_9 w_s^2 T - 2c_{10} w_s T^2\end{aligned} \tag{6.31}$$

In the case of seawater desalination, the theoretical minimum energy for separation is shown in Figure 6.2 by changing the seawater concentration and temperature. The theoretical minimum energy was calculated using method 2. The theoretical minimum energy for the separation of fresh water from seawater increases with increasing seawater concentration and temperature. It can be easily found that the effect of concentration on the theoretical minimum energy for separation is much larger than the effect of temperature. Thus, seawater concentration is the most important variable for estimating the

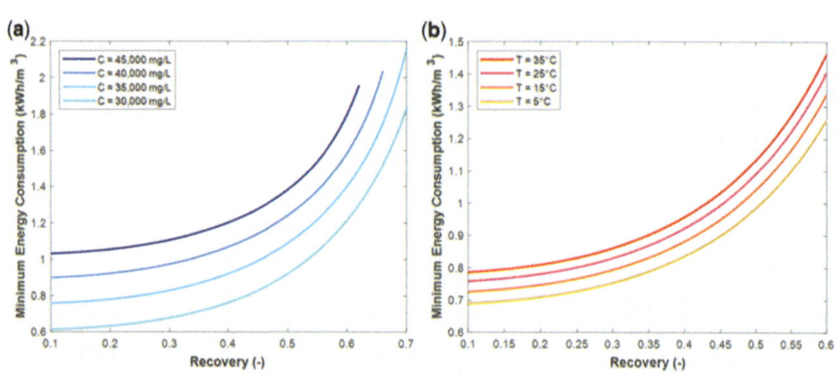

Figure 6.2 Theoretical minimum energy consumption calculated using method 2 for seawater desalination: (a) the effect of seawater concentration at 25°C, and (b) the effect of seawater temperature at 35 000 mg/L.

theoretical minimum energy required for separation in seawater desalination. However, the temperature of the seawater is significantly affected by seasonal temperature variation, and the effect of temperature on the energy consumption of the SWRO process cannot be underestimated.

If the concentration and temperature of seawater are 35 000 mg/L and 25°C, respectively, the theoretical minimum energy of seawater desalination is 1.07 kWh/m^3 for 50% recovery [16, 36, 45, 305]. However, the actual energy consumption of the SWRO process is approximately 2–4 kWh/m^3, which is much higher than the theoretical minimum energy required for separation. This reveals that numerous non-ideal conditions influence the higher energy consumption of the SWRO process. These numerous non-ideal conditions are the key factors for reducing the actual energy consumption in the SWRO process. In addition, the additional energy consumption compared to the theoretical minimum energy for separation is approximately 1–3 kWh/m^3, so the energy efficiency of the SWRO process might be improved further. To reduce the actual energy consumption of the SWRO process, the main factors arising from additional energy consumption should be examined.

6.1.2 Analysis of future directions for low-energy SWRO

The simulation approach is useful for identifying the additional energy consumption in each unit compared to the theoretical minimum energy required for separation. In this section, one example is selected to analyze the current limitations and future directions for improving the energy efficiency of the SWRO process. A feed solution with 35 000 mg/L concentration, 25°C temperature, 1 bar pressure, and 20 000 m^3/d flow rate is fed into the SWRO system. It is assumed that only NaCl is the solute species dissolved in the feed solution. In the SWRO system example, HPP, BP, and ERD are installed to describe the current standard configuration of the SWRO. The recovery of the example is assumed to be 50%, so the permeate production rate is 10 000 m^3/d. The theoretical minimum energy for separation by SWRO can then be calculated from the exergy analysis, and the actual energy consumption in each energy-consuming unit is displayed in Figure 6.3. An SWRO model including mass balance, flux equations, pressure drop, and concentration polarization is utilized to estimate the actual energy consumption of each unit in the SWRO system [17, 33, 157]. The design parameters and variables are listed in Table 6.2. The actual energy consumption is 2.26 kWh/m^3 by including 2.07 kWh/m^3 from HPP and 0.19 kWh/m^3 from BP. The theoretical minimum energy for separation is obtained from the difference between the feed stream and outlet streams (permeate and retentate). Despite the minimum energy consumption (1.07 kWh/m^3), the actual energy consumption is 2.26 kWh/m^3, so the exergy efficiency of the SWRO is 47.35%. This result reveals that additional energy of 1.19 kWh/m^3 is needed to operate the actual SWRO system. Thus, a solution for improving the energy efficiency of SWRO can be obtained from the analysis of the additional energy consumption.

The breakdown data for the actual energy consumption are shown in Figure 6.4. The factors that contribute to the additional energy consumption of the SWRO process can be identified based on these data. Even though these data

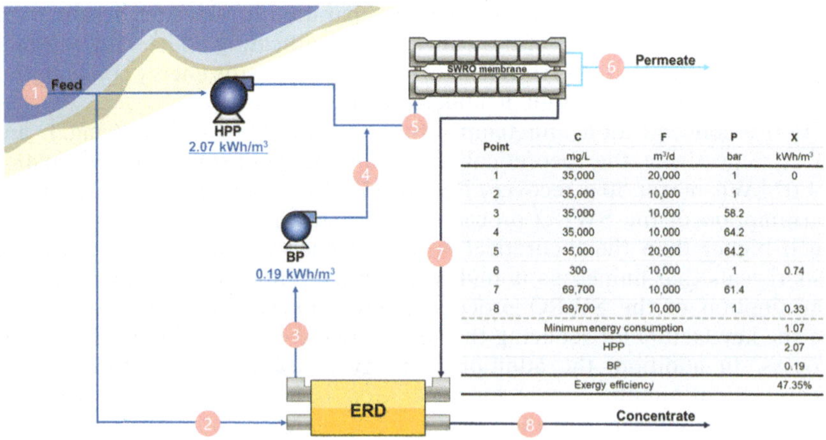

Figure 6.3 Exergy analysis and actual energy consumption of the SWRO process. C is the concentration, T is the temperature, F is the volumetric flow rate, P is the pressure, and X is the exergy. T is fixed at 25°C.

Table 6.2 SWRO process variables and parameters in this case study.

Name	Value	References
Pump efficiency	85%	[306, 307]
ERD efficiency	95%	[306, 307]
Feed concentration	35 000 mg/L	[17, 21]
Feed temperature	25°C	[308]
Feed flow rate	20 000 m³/d	
Number of RO module in series	7	[309]
Total recovery	50%	[21]
RO module	Dow Chemical FILMTEC SW30XLE-400	[310, 311]

correspond to only one example, the data trend is useful for investigating future directions for the improvement of energy efficiency in SWRO. The most significant factor for the additional SEC of the SWRO process is the irreversibility of the HPP (0.69 kWh/m³). Owing to the inefficiency of the HPP and BP, the process requires the addition of 0.31 and 0.19 kWh/m³, respectively. It is not easy to reduce further these two additional energy consumptions caused by the inefficiency, because the process equipment (HPP, ERD, and RO modules) has already reached its practical maximum efficiencies. The pump and ERD efficiencies in this simulation are 85% and 95%, respectively. Although some reports have shown the possibility of very high efficiency in the HPP and ERD (90% of pump and 98% of ERD) [36, 95], such high-performance equipment is usually expensive and the actual benefit of using it is not significant. Therefore,

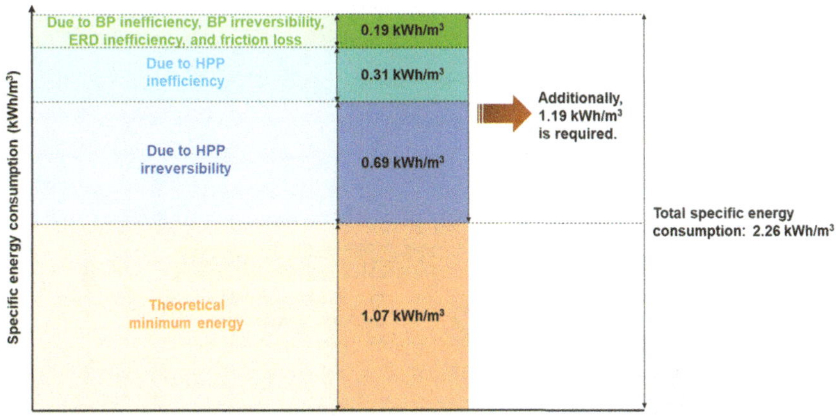

Figure 6.4 Actual energy consumption of the SWRO process and its breakdown data according to the causes of energy consumption.

in this study, factors that can be overcome by improving the efficiency of the equipment itself were not considered.

Then, two factors (theoretical minimum energy and irreversibility of HPP) remained and were considered to identify ways to reduce additional energy consumption in the SWRO process. In addition, high osmotic energy in the RO retentate can be recovered to reduce the overall energy consumption of the SWRO system. Thus, three improvement axes for the low-energy SWRO process can be categorized as follows: (a) minimizing the irreversibility of HPP, (b) decreasing the osmotic pressure of seawater, and (c) recovering osmotic energy from the RO retentate. To analyze the maximum potential of these improvement axes, the maximum available energy margin in each axis should be estimated. The extent of improvement in each axis by applying many approaches for reducing the SEC of the SWRO process should be reviewed extensively and analyzed critically.

These axes are graphically represented by a situation in which a runner tries to jump a hurdle, as shown in Figure 6.5. The objective of the runner is to jump the hurdle using the runner's minimum energy. The hurdle height is the minimum jump height that should be overcome by the runner. Thus, the hurdle height is explained by the theoretical minimum energy required for separation, which correlates with the osmotic pressure of seawater. If the runner jumps too high over the hurdle, the runner wastes energy. The additional jump height is correlated with the irreversibility of the HPP. Finally, the jump energy required to overcome the hurdle is reduced if the runner can place a steppingstone on the jump track. In other words, the osmotic energy recovery from the RO retentate helps reduce the overall SEC of the SWRO in the same way as placing a steppingstone on the jump track. In summary, these improvement axes are helpful in reducing the overall SEC in the SWRO, and the strategies are easily understood using the example of a runner jumping a hurdle.

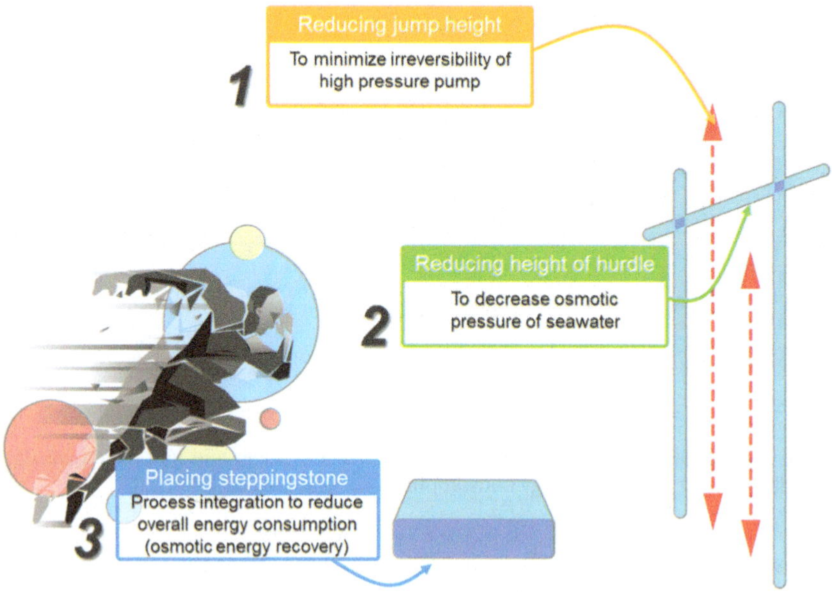

Figure 6.5 Graphical representation of the improvement axes for a low-energy SWRO process.

6.1.3 Maximum available margin for low-energy SWRO

From the listed improvement axes, the amount of energy consumption that would be reduced in each improvement axis should be analyzed. Thus, the maximum available margins for each improvement axis are estimated from the exergy analysis to identify the amount of energy consumed by the SWRO process that could be reduced theoretically. The maximum available margins for each axis were calculated from the actual energy consumption of the example SWRO process in the previous section, and the results are shown in Table 6.3. The additional feed solution, which has a lower concentration, is required to recover the osmotic energy in the concentrate. The amount of energy recovered from the concentrate is a function of the concentration and flow rate of the additional feed solution [125, 140, 312, 313]. Therefore, to calculate the maximum available margin for axis (c), the concentration and

Table 6.3 Maximum available margin for each improvement axis to reduce SEC in the SWRO process.

Axis for reducing energy consumption	Maximum available margin
(a) Minimize irreversibility in HPPs	0.69 kWh/m^3
(b) Decrease osmotic pressure of seawater	1.07 kWh/m^3
(c) Recover osmotic energy from RO retentate (2000 mg/L and 20 000 m^3/d brackish water basis)	0.21–0.56 kWh/m^3 (20–60% brine dilution)

flow rate of the additional feed solution should be determined. In this case study, it was assumed that a brackish water feed solution with a 2000 mg/L NaCl concentration and 20 000 m^3/d flow rate is fed into the SWRO system to recover the osmotic energy of the retentate. By changing the ratio of dilution in the retentate after an osmotic energy recovery process, such as pressure-retarded osmosis (PRO) or reverse electrodialysis (RED), the exergy analysis was utilized to estimate the maximum obtainable energy.

Among the improvement axes, axis (b) shows the largest margin, and axis (c) exhibits the lowest available margin for reducing SEC in the SWRO process. However, the maximum margin exists only for the possibility of improving energy efficiency. The actual achievable energy reduction for each axis may differ depending on the difficulty of the application method. Therefore, state-of-the-art technologies for reducing SEC in the SWRO process according to each improvement axis should be analyzed. Then, the method most suitable for achieving a low-energy SWRO process should be identified, and the extent to which the SEC of the SWRO process can be reduced by applying state-of-the-art technologies should be estimated.

6.2 MINIMIZING IRREVERSIBILITY IN A HIGH-PRESSURE PUMP

To analyze the irreversibility in a high-pressure pump, it is assumed that the seawater feed solution is pretreated, pressurized, and supplied to an RO membrane system in the SWRO system. The pressurization system consists of HPP and/or BP, followed by ERD. The pressure energy supplied to the SWRO system is utilized to obtain pure water from the seawater feed solution. However, an additional different type of work that is not directly involved in separation is also associated with the process during pressurization. This work is correlated with the energy lost by entropy generation [36, 314]. In other words, the separation of seawater can be classified into two parts: reversible work (the theoretical minimum energy for separation) and irreversible work (energy lost by entropy generation). To minimize the irreversible work in the SWRO process, it might be useful to supply pressure in a staged manner rather than all at once [11, 12, 21]. For this purpose, staged designs have been developed to increase the hydraulic pressure gradually, and the number of stages is determined by situations such as recovery and feed concentration.

6.2.1 Multistage RO

The design configurations for multistage RO are summarized in Figure 6.6. A multi-stage RO system adopts several stages of RO operation, where the retentate resulting from the previous stage becomes the feed stream to the next stage (Figure 6.6a). The feed pressure of the retentate from the previous stage should be increased by the inter-stage BPs before supplying the next stage to satisfy the operating pressure required in the next stage. After the final stage, the ERD recovers the remaining pressure energy in the final retentate to maximize energy efficiency. At very high recoveries, such as multi-stage brackish water reverse osmosis (BWRO), BPs are not needed before the next stage because the hydraulic pressure applied by the HPP is sufficiently high. Theoretically,

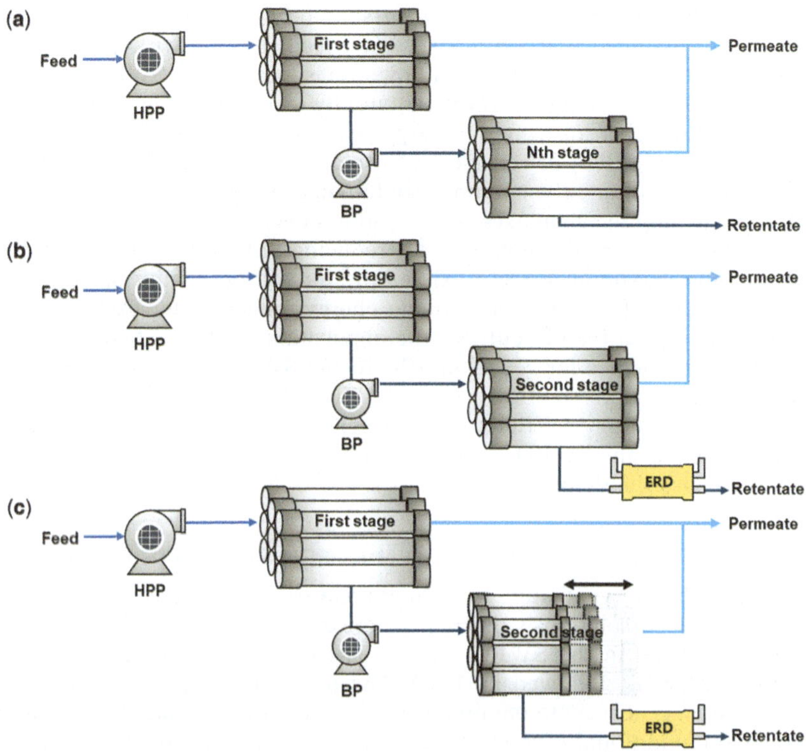

Figure 6.6 RO staged designs for minimizing the irreversible work of HPP: (a) multi-stage, (b) brine conversion system (BCS), (c) low-pressure multi-stage system (LMS).

infinite multi-stage RO is preferred to minimize irreversible work (energy loss). However, this is not practical because the capital cost of an infinite stage system becomes tremendous [21]. The practical limitation of a multistage RO application in more than three stages has already been mentioned in another study [45]. The current research points out that a two-stage operation is regarded as one of the most plausible designs in the multistage RO system for SWRO.

Although the benefit of the multistage RO design is attractive for low-energy SWRO systems, the single-pass SWRO process is usually designed using a single-stage system [11]. Owing to the high osmotic pressure of the retentate from the first stage, the pressure required before the second stage should be very high. Thus, desalination equipment such as RO membranes and pressure vessels in the second stage should be more resistant to high-pressure operation [315]. Although high-pressure operation is not desirable, some desalination plants implement two-stage SWRO configurations with a brine conversion system (BCS) to increase the recovery of the SWRO process (Figure 6.6b) [180, 315]. In the BCS system, another stage is added sequentially after the conventional single-stage SWRO system. The second stage is used to produce more permeates.

The retentate from the first stage is pressurized and fed into the BCS system. The advantage of the BCS system is that it can reduce the size of the RO plant installation space. In addition, owing to its enhanced recovery, the volume of the brine discharged into the ocean can be reduced. However, a high SEC is inevitable owing to the high recovery [286]. Recently, a low-pressure multi-stage system (LMS) was developed by modifying the original configuration of the conventional single-stage RO (with one PV) as a multi-stage RO (two or three PVs) (Figure 6.6c) [286, 297]. In the LMS system, only two to four RO modules are included in the modified PV of the first stage, unlike seven or eight elements included in the conventional single-stage RO. More than four RO modules are contained in the typical PVs in the second stage. It was reported that the energy consumption would be reduced by 20% compared to the conventional single-stage RO owing to its high recovery [297]. However, it should be noted that current SWRO desalination plants with multistage RO are not targeted at such low energy consumption, because the multistage configuration is usually designed to reach a high-recovery desalination system [17, 297]. If the multistage SWRO system is operated with high recovery (Y: recovery, $Y > 50\%$), it is inevitable to cost higher SEC than an SWRO system with typical recovery ($Y < 50\%$). This is contrary to the theoretical understanding of multistage RO systems discussed in the previous chapters. This is because the basis of recovery for comparison in SEC is different in conventional single-stage RO and multistage RO systems. Thus, the SEC of single-stage and multistage RO should be compared at the same recovery to validate the claim that multistage RO can effectively reduce the irreversibility in the HPP.

Although energy efficiency might be improved by the utilization of the multistage RO system, the increased capital cost and process robustness at high pressure should be considered seriously. A critical issue is that if the ERD is placed between the retentate after the second stage and the feed stream before the first stage to recover the pressure energy in the retentate after the second stage, the pressure of the seawater feed after the ERD is higher than the operating pressure required in the first stage RO. Then, the process cannot be operated owing to the imbalance of the pressure [316]. Two remedies are proposed to solve this issue. The first solution is that the ERD should be placed between the retentate streams from the first and second stages to recover the pressure energy from the retentate stream of the second stage. However, this solution has a disadvantage in that it cannot recover the entire pressure energy in the retentate stream from the second stage using only a single ERD. Therefore, an additional ERD should be installed between the feed stream to the first stage and the retentate stream after the first ERD to recover the pressure energy of the retentate stream entirely. The disadvantages of the first solution are the rise of ERD costs and the reduction of the overall energy efficiency of the ERD due to the series implementation of the two ERDs. The second solution is the employment of turbine ERDs [such as FT and Pelton turbine (PT)] instead of isobaric ERDs [such as PX and dual work exchanger energy recovery (DWEER)]. However, turbine ERDs usually have lower energy efficiency than isobaric ERDs. These disadvantages should be considered in the final decision-making process, even if the energy consumption of staged RO designs can be reduced.

Recently, two-stage RO systems have been investigated in terms of energy efficiency and practical applicability [103, 104, 317–319]. The simulation results revealed that two-stage SWRO can be more energy-efficient than a single-stage SWRO when the recovery is higher than 50% [104]. However, at a low recovery of less than 45%, the single-stage RO might be more energy-efficient than the two-stage RO, even though the staged RO usually has low irreversible energy loss in the high-pressure pump. Thus, the two-stage RO should be implemented by considering the practical energy analysis with module-scale simulation and optimization, not only considering theoretical intuition. Kim *et al.* reported that the types of ERDs utilized in a two-stage RO system would differ depending on the target recovery [104]. A two-stage RO can be effectively implemented for the treatment of highly saline, high-temperature seawater [103]. The ISD and SPSP design configurations can also be applied in a two-stage RO system to further improve the energy efficiency. Therefore, the staged RO configuration has high potential for improving the energy efficiency, especially in high-recovery desalination or high-salinity and high-temperature water treatment systems.

6.2.2 Semi-batch RO

It is impractical to install an infinite multistage RO system. A batch-type reactor strategy can be adopted in the RO configuration to overcome the installation difficulty of the infinite multistage RO. A semi-batch RO called closed-circuit desalination (CCD) or closed-circuit reverse osmosis (CCRO), has been proposed to realize an infinite multistage RO system in the form of a batch configuration [320]. The schematic diagram of the semi-batch RO is illustrated in Figure 6.7. The CCD or semi-batch RO configuration adopts a cyclical process in which the retentate from the RO module is continuously recycled as feed and mixed with fresh feed. The concentration in the semi-batch RO increases continuously with the operation time taken from the start of the semi-batch RO operation, because the salt in the feed solution cannot escape easily from the semi-batch RO system, unlike the water. The cyclic operation is terminated if the desired recovery has been reached. Then, the concentrated solution in the semi-batch RO system is purged to start a new cycle. In the system, the hydraulic pressure of the retentate should be boosted by a circulation pump (CP) and supplied to

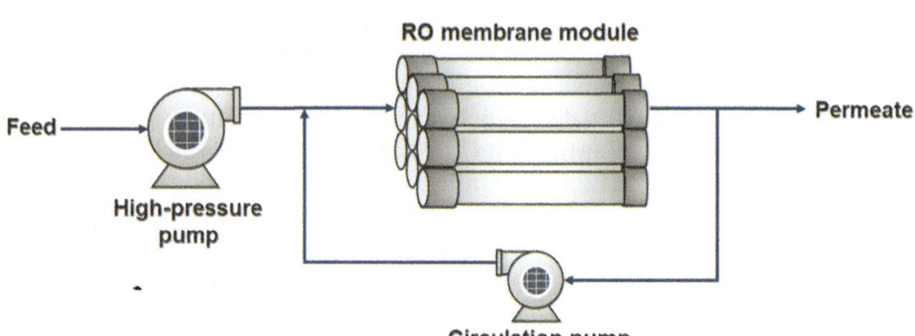

Figure 6.7 Semi-batch RO design for minimizing the irreversible work of HPP.

the RO system without using ERDs. Instead, variable pressure is applied using HPP as the concentration in the semi-batch RO system increases to produce permeate water flux consistently during each cycle.

It has been claimed that the semi-batch RO process has several advantages over the conventional RO system, such as significant reduction in energy consumption, higher product recovery, higher resistance to fouling and scaling, and optimum flexibility in operation [321]. In addition, a semi-batch RO can simulate the effect of a multistage RO configuration without installing an actual multistage RO system. Thus, a semi-batch RO can also reduce the capital cost compared to the multistage RO configuration.

Numerous studies have been conducted to investigate the actual applicability and practical energy efficiency of the semi-batch RO process [99, 322–327]. As a conventional BWRO with high recovery does not include ERD due to high capital cost and low return of investment, the semi-batch RO is competitive in terms of energy efficiency for brackish water desalination at high recovery [325, 326]. However, the continuous mixing between the recycled feed stream and fresh feed has a negative influence on energy efficiency [12, 97]. Therefore, the applicability of semi-batch RO for seawater desalination has not been justified. Until now, the semi-batch RO method has been applied to systems for the treatment of brackish water [99]. Although a pilot semi-batch RO has been tested with a seawater feed, a full-scale semi-batch RO for SWRO has not yet been demonstrated.

6.2.3 Batch RO

The batch RO system has been recently suggested as a prominent candidate for minimizing the irreversibility associated with the high pressures used in the desalination process (Figure 6.8) [322, 328–333]. Conventional one-stage continuous RO applies constant pressure to the feed solution to obtain sufficient driving force for water to permeate the RO membrane. However, this constant pressure operation has an inevitable problem of irreversible exergy loss in the membrane module, leading to excessive power consumption by the high-pressure pump. Batch RO can minimize the irreversible exergy loss by gradually increasing the pressure according to the changing feed concentration.

In particular, if high recovery is required, the benefit of employing batch RO will be maximized by achieving high recovery with low energy input. With this advantage of high energy efficiency, the applicability and feasibility of batch RO systems have been investigated extensively. In particular, in processes that usually involve high recovery operations such as groundwater and BWRO, the advantage of batch RO would be maximized because a large amount of irreversible energy loss cannot be avoided in a continuous RO operation with high recovery. In other words, the batch RO process can be effectively utilized for groundwater and brackish water treatment systems [322, 328, 334]. This has led many researchers to investigate how to apply the batch RO system effectively to brackish water and groundwater treatment systems [328, 330–332, 334]. A recent study on the batch RO system for brackish water desalination revealed that the actual energy consumption with a pilot-scale system (>10 m^3/d) is approximately 0.39 kWh/m^3, and the second law efficiency is 33.2% [98].

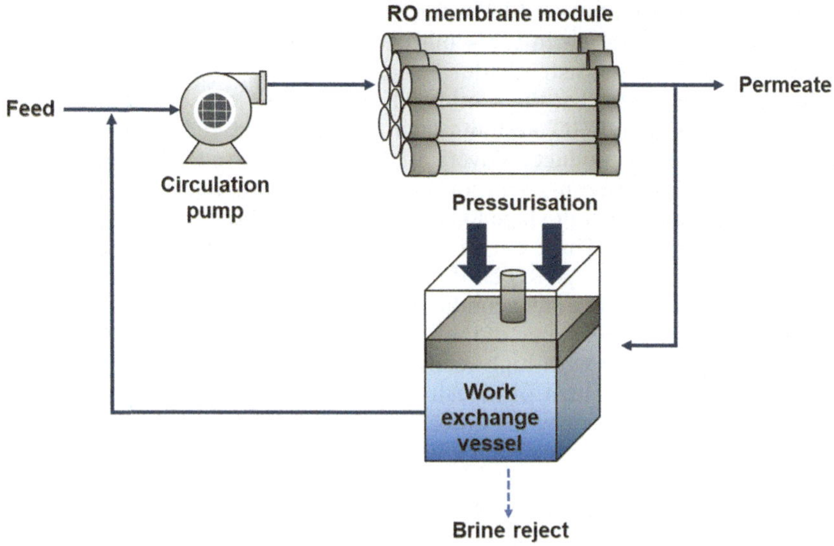

Figure 6.8 Batch RO design for minimizing the irreversible work of HPP.

The existing BWRO systems exhibit SEC in the range of 0.5–1 kWh/m^3, and the second law efficiency is approximately 10–15% [335]. The high energy efficiency of batch RO is attributed to the minimization of irreversibility during pressurization.

Recently, batch RO has been implemented in seawater desalination to maximize energy efficiency [336–340]. As seawater desalination has the largest capacity and demand worldwide, it is attractive to develop a new energy-efficient desalination system to overcome the current energy level of the continuous RO system. It is expected that the batch RO for seawater desalination can save 11% of the SEC compared to the conventional SWRO at 35 000 mg/L salinity and 50% recovery [338, 339]. These studies demonstrated that the energy efficiency of batch RO can be improved further, and the cost of seawater desalination can be reduced.

Although the theoretical advantages of batch RO are acknowledged, its actual implementation is quite challenging, especially in large-scale systems [322]. Some researchers have investigated the actual applicability and feasibility of batch RO [98, 322, 323, 331, 333, 341–343]. Two types of batch RO designs, that is, free-piston [332, 344] and flexible bladder [339, 345, 346], were developed for practical operation and realization. These studies validated experimentally the batch RO system [332, 344–347], and identified non-ideal factors including salt retention factor and longitudinal dispersion in the batch RO [98, 330, 332], while modeling them theoretically [322, 323, 333, 339]. Recently, researchers investigated the extension of the application of the batch RO for large-scale operations involving inland brackish water desalination [98] and full-scale systems for seawater desalination [339, 346].

To advance batch RO technology, recent studies have focused on model development, providing promising predictions with regard to inorganic fouling and energy efficiency [323, 333, 348]. However, the correlation between flow distribution and fouling deposition has not been investigated, so a comprehensive analysis considering mass transfer, flow distribution, fouling deposition, energy, and cost estimation has not been conducted because of the lack of a fully validated study covering all key aspects of batch RO. In addition, there is a lack of knowledge about process design, optimization, and testing correlated to real batch RO systems. The lack of such knowledge about batch RO limits its actual application as an efficient desalination technology in groundwater treatment systems, which is the main research gap. Thus, these will be the main issues researchers should focus on for the development of batch RO systems for desalination technology.

6.2.4 Hybrid batch RO

Recently, emphasis has been placed to high-recovery desalination systems in efforts to reduce ecological and environmental impacts [27, 349]. Zero liquid discharge (ZLD) is a prominent remediation method for the treatment of concentrated brine from desalination systems without ecological and environmental impacts [350]. However, building a complete ZLD system is quite challenging because of the intensive use of energy and cost [350–352]. Recently, minimal liquid discharge (MLD) has been proposed as an alternative to ZLD, not only to avoid the huge amounts of energy and cost but also to obtain ecological and environmental benefits [22, 353]. Unlike ZLD, in MLD, membrane-based desalination processes can be employed. In other words, high-efficiency desalination processes such as RO can be implemented to achieve MLD. Novel RO-based processes including staged RO [104, 354], low-salt-rejection reverse osmosis (LSSRO) [353], osmotically assisted RO (OARO) [22, 355], and draw solution-assisted RO (DSARO) [33] have been suggested for MLD. However, these alternative processes are based on a continuous RO operation, so the energy efficiency is limited by the characteristics of the continuous RO, such as irreversible exergy loss in instantaneous pressurization [11, 12, 21].

In the MLD system, batch RO is a favorable process for achieving high recovery and high energy efficiency. As was shown in Section 6.2.3, numerous studies on batch RO have focused on filling the research gap between conceptual design and actual realization [98, 333, 338, 339, 341]. These studies revealed that a large work exchange vessel (in the free-piston design) or flexible bladder to achieve a high recovery of over 0.8 are inevitable problems in a batch RO system [98, 339, 346]. The large size of the work exchange vessel not only increases the capital cost of the batch RO system, but also increases the length of pipes, which could increase the effect of salt retention and the pressure drop along the pipes. It is desirable to reduce the size of the work exchange vessel in a high-recovery desalination system using batch RO, if possible.

Semi-batch RO is free of the problem of large work exchange vessels. However, semi-batch RO cannot avoid the problem of additional entropy generation. The advantages and disadvantages of semi-batch RO are opposite to those of batch RO. Thus, these characteristics imply that batch RO and semi-batch RO could complement each other in terms of the large work exchange vessels and energy minimization problems. If a new hybrid process incorporating batch RO and semi-batch RO could be developed, the hybrid process would be utilized for a high-recovery desalination system without excessive energy consumption, unlike the semi-batch RO and too large a work exchange vessel, unlike the batch RO. In other words, the hybrid process could be a compact and energy-efficient desalination system.

The hybrid batch/semi-batch RO (HBSRO) process is proposed to achieve high recovery and a highly energy-efficient desalination system with a compact size of work exchange vessels [97]. HBSRO is driven by integrating the operating strategies of batch RO (Figure 6.9a) and semi-batch RO (Figure 6.9b) systems. As the HBSRO process is designed based on batch operation, it is a cyclic system in which a single operation cycle is repeated periodically. At the earlier stage of

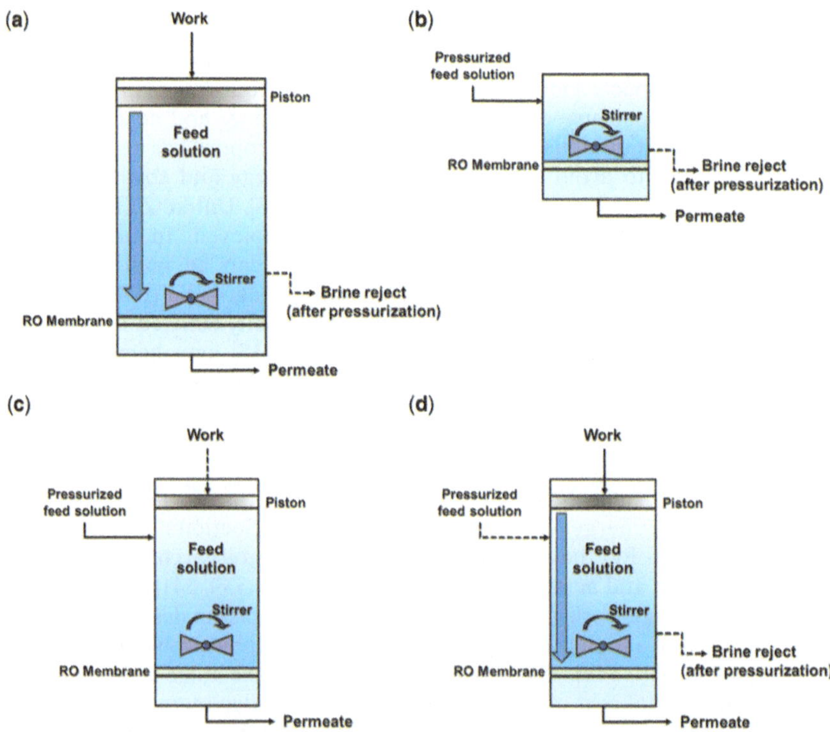

Figure 6.9 Schematic diagram of working principles in (a) batch RO, (b) semi-batch RO, (c) HBSRO (semi-batch phase), and (d) HBSRO (batch phase). The solid line and dashed line denote the active line and inactive line, respectively. The relative size of the vessels denotes that batch RO has the largest size cell, followed by HBSRO, and semi-batch RO has the smallest.

each cycle (semi-batch phase), the semi-batch RO operation is employed in the HBSRO process (Figure 6.9c). The pressurized feed solution (higher than the osmotic pressure of the feed) is fed into the membrane cell to produce permeate water across the RO membrane. The concentration inside the membrane cell increases with the retention of salt during the semi-batch phase. When the concentration reaches a certain level, the operation phase shifts from the semi-batch phase to the batch phase. In the batch phase, a piston in the membrane cell travels toward the RO membrane by work, and the water in the cell permeates the RO membrane (Figure 6.9d). After the piston reaches a certain position, the concentrated brine remaining in the cell is purged out, and a new feed solution is refilled in the cell. Then, the batch RO phase is completed. Through these working principles, a single cycle of the HBSRO process is completed.

The rationale of the HBSRO process is that it can make a compact size of the membrane cell even at high recovery. In the existing practical batch RO systems, the conceptual membrane cell was realized by connecting a commercial RO module (such as a spiral wound type) with a work exchange vessel (in free-piston design) or a bladder (in flexible bladder design) to supply the work required for batch RO [98, 345, 346]. At the start of the batch RO cycle, the feed solution is placed in the work exchange vessel (or flexible bladder) and the RO module. After pressurization (by pushing the free-piston or shrinking the flexible bladder), the feed solution in the batch RO system permeates the RO membrane, and the amount of permeate in each batch RO cycle is the same as the volume of the work exchange vessel (or flexible bladder) [98, 345, 346]. Meanwhile, the concentrated brine remains in the RO module and is purged for the next cycle. The recovery of the batch RO is calculated by the volume of the work exchange vessel (or flexible bladder) over the summation of the volume of the RO module and the work exchange vessel (if the volume of the connecting pipes is neglected) [98]. Because the volume of the commercial RO module is fixed, the work exchange vessel (or flexible bladder) should have a very large volume to achieve high recovery. For example, a pilot-scale, free-piston-type batch RO system with an 8-inch RO module requires 64.6 and 145 L of work exchange vessel to reach 0.8 and 0.9 recoveries, respectively [98]. If these volumes would be prepared with a commercial 8-inch pressure exchange vessel, and the length of the 8-inch vessel should be approximately 2 m for the recovery of 0.8 and 4.5 m for the recovery of 0.9. The long pressure exchange vessel requires a longer length of pipes connecting the pressure exchange vessel and the RO module. As reported in our previous paper, the increased pipe volume caused a significant effect of salt retention inside the pipes [98]. Thus, the length of the pressure-exchange vessel should not be too long. However, a pressure exchange vessel with a diameter larger than 8 inches is not widely available commercially, because a large-size RO module over 8 inches is under development for application in actual RO plants [356, 357]. In addition, the very large volume of the work exchanger vessel might have a significant non-ideal mixing inside the vessel, which has a detrimental influence on the SEC.

In the HBSRO process, the required volume of the work exchange vessel can be reduced by adapting the semi-batch RO phase before the batch RO operation. At the start of the HBSRO process cycle, the feed stream is fed into the HBSRO process in the semi-batch RO phase. The feed solution is fed from the outside of the HBSRO system, and not from the work exchange vessel, unlike the conventional batch RO. The energy penalty arising from the semi-batch RO operation is minimized owing to the existence of the work exchange vessel. Unlike the conventional semi-batch RO configuration, the proposed HBSRO process includes a work-exchange vessel. The entire volume of the HBSRO process, including the RO module, work exchange vessel, and pipes, is much larger than the volume of the conventional semi-batch RO process (including only the RO module and pipes). Thus, the concentration increase in the HBSRO process during the semi-batch RO operation is not rapid, unlike the conventional semi-batch RO, so the entropy generation by mixing in the HBSRO process is lower than that in the conventional semi-batch RO. This is the most remarkable feature of the HBSRO process. Compared to the conventional batch RO, the HBSRO can realize the compact size of the pressure exchange vessel, compromising the minimum energy penalty by using the semi-batch RO operation.

Figure 6.10 shows the theoretical minimum pressure (osmotic pressure) in the semi-batch RO, batch RO, and HBSRO processes. The area below each line

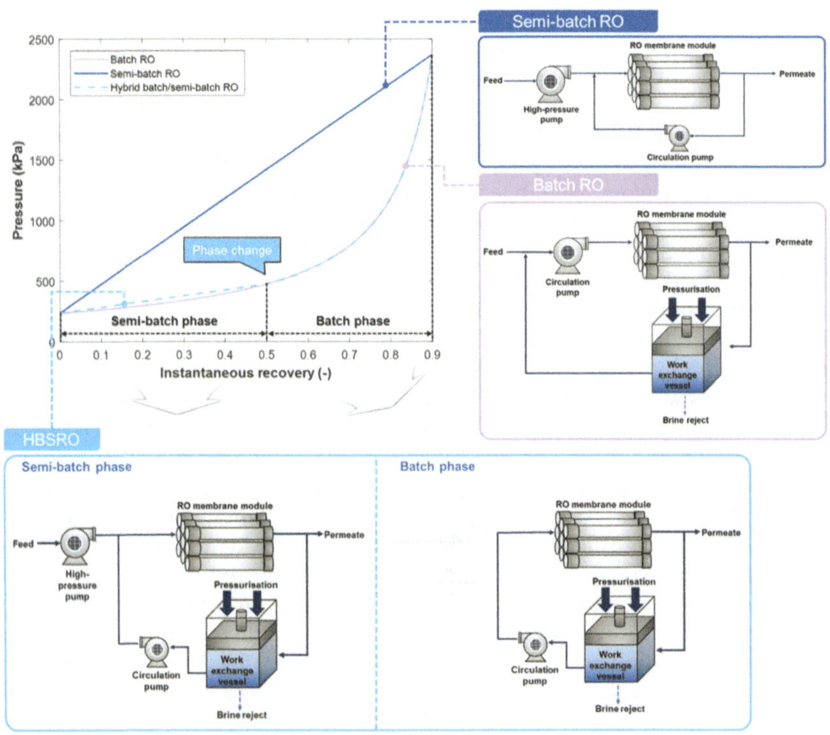

Figure 6.10 Theoretical rationale of the HBSRO compared with the semi-batch RO and the batch RO.

denotes SEC. The osmotic pressure profile of the semi-batch RO is linear, and that of the batch RO is similar to an exponential function [323, 331]. In the case of HBSRO, the linear pressure profile in the semi-batch phase has a low slope owing to the existence of the work exchange vessel. The large system volume, including the work exchange vessel, leads to a slow increase in concentration in the system under the same permeate flow rate. Thus, the energy penalty in the HBSRO can be minimized. After finishing the semi-batch phase, the sequential batch phase follows in the entire HBSRO operation. The HBSRO then makes the minimum energy consumption theoretically during the batch phase by following the pressure profile of the batch RO. As shown in Figure 6.10, the energy penalty in HBSRO is determined by the proportion of the semi-batch phase to the total operation time of the HBSRO. There is a trade-off between the energy penalty and the reduced volume of the work exchange vessel, and each benefit and loss can be determined by the proportion of the semi-batch phase.

A recent study revealed that a HBSRO with a work exchanger volume half that of batch RO yields a lower SEC than batch RO due to energy saving in the purge-and-refill phase in a practical design working above a recovery of 0.9 [97]. This advantage becomes significant with an increase in recovery. In other words, the HBSRO can effectively reduce the size of the work exchange vessel and also obtain low-energy consumption if desalination with very high recovery (>0.9) should be conducted. Thus, HBSRO is a competitive RO configuration with MLD and ZLD systems.

6.2.5 Energy comparison of minimizing irreversibility in HPP

It has been claimed that staged RO configurations can reduce SEC compared to single-stage RO [45]. However, this claim has not been clearly demonstrated in numerous studies, because the staged RO configuration is usually utilized to apply a high-recovery desalination system rather than energy minimization [101, 323, 327]. Thus, the SEC reduction in the staged RO configuration should be validated by comparison under equivalent conditions. In some studies, a comparative analysis was performed by simulating the change in SEC at each stage of the RO design according to the overall recovery [103, 104, 317, 318, 322].

The comparative analysis demonstrated that batch RO, which is ideally operated with variable feed volume, consumes the lowest energy in all ranges of recovery among all staged RO configurations under ideal conditions. The minimum irreversibility in the batch RO system is the main reason for the minimum energy consumption. However, semi-batch RO shows higher SEC than two-stage RO in typical SWRO recovery (40–55%) [41, 322]. Continuous entropy generation in semi-batch RO, which occurs when fresh feed is mixed with retentate, is the most critical issue in SEC minimization. Furthermore, the SEC of semi-batch RO can deteriorate as the ERD efficiency decreases. Thus, semi-batch RO is not an energy-efficient configuration under the condition of typical SWRO. In addition, it has been reported that a two-stage RO configuration consumes less energy than semi-batch RO, but more energy than ideal batch RO [41, 322]. Overall, it can be concluded that staged RO designs consume less energy than conventional single-stage RO designs.

However, the SEC shows different results when considering a practical situation. It should be noted that a modified batch RO designed to operate in a practical situation cannot reduce the SEC effectively compared to the conventional single-stage SWRO process. Owing to the utilization of ERD, ERD inefficiency influences the energy lost in each cycle. As a result, the modified batch RO consumes more energy than the single-pass RO in a recovery range of less than 45% [322]. Even at a recovery of over 45%, the SEC of the modified batch RO is still higher than that of the semi-batch RO or two-stage RO. The pressure loss caused by continuous circulation in the modified batch RO is higher than that in the semi-batch RO and two-stage RO configurations. Therefore, the modified batch RO configuration is not energy efficient. To maximize the potential of the batch RO configuration, a more energy-efficient configuration that is close to the ideal batch RO should be developed.

The current development in the batch RO configurations is towards the realization of the ideal batch RO system. A flexible-bladder-type batch RO was designed to simulate the batch RO implementation for seawater desalination [338, 339]. A free-piston type was implemented as a pilot-scale system to extend the applicability to a large-scale desalination system [97, 98]. Thus, it is unnecessary to depend on the modified batch RO to realize the batch RO configuration in the practical situation. Although the comparative analysis between the ideal batch RO and the practical batch RO systems such as the free-piston and flexible-bladder types has not been suggested, it is desirable to note the energy efficiency of batch RO configuration in near future.

In the current situation, two-stage RO is the most energy-efficient RO configuration for practical operation, along with the improvement strategy of minimizing the irreversibility in HPP. To compare the energy state in the current situation, the changes in the flow exergy and the extent of irreversibility in the staged RO configurations are shown in Figure 6.11. Although ideal

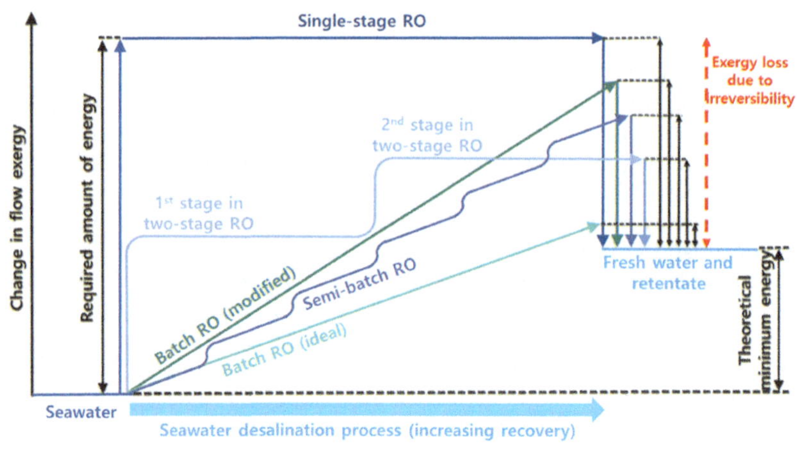

Figure 6.11 Staged RO configuration and flow exergy analysis in seawater desalination.

batch RO may be the most energy-efficient, two-stage RO can be regarded as a practical solution for minimizing the irreversibility in HPP. If the two-stage RO configuration instead of single-stage RO were successfully applied, this would reduce energy consumption to the greatest extent: approximately 0.34 kWh/m^3 of energy reduction compared to the single-stage RO, while the semi-batch RO and the modified batch RO exhibit an energy reduction of 0.29 and 0.13 kWh/m^3, respectively, with a moderate circulation pressure loss of 0.1 bar [322]. The energy reduction in the staged RO configuration is estimated from the case study [322]. However, the energy reduction is a good barometer in the current situation of effectiveness among staged RO configurations.

6.3 REDUCING THE OSMOTIC PRESSURE OF SEAWATER

SWRO desalination plants produce pure water by applying a hydraulic pressure higher than the feed osmotic pressure. If seawater is blended with a low-salinity stream (e.g., domestic wastewater, river water, and so on) or the permeate concentration is adjusted to increase, the osmotic pressure gradient between the feed water and permeate will decrease. This can reduce the applied hydraulic pressure, thereby resulting in a lower energy requirement for the production of pure water. To reduce the osmotic pressure gradient, three technologies have been proposed: (a) a split partial single-pass (SSP) RO system, (b) a forward osmosis (FO) and RO hybrid system, and (c) a draw solution-assisted RO (DSARO) system. This section discusses these low-energy technologies for SWRO desalination.

6.3.1 Split partial single pass

Single-pass RO is the most common configuration used in seawater desalination because of its simple configuration. To enhance its performance, an internally staged design (ISD) was developed [358]. In ISD, a high-rejection membrane and a high-flux membrane are located at the head and tail, respectively, of the PV for a uniform flux distribution inside the PV. This can reduce SEC and increase the lifetime of the membranes, while the water quality of its product water can be lower because of the high-flux membranes at the rear of the PV. Therefore, a split partial single-pass (SSP) RO system, as depicted in Figure 6.12(a), was developed by combining a single-pass RO design with ISD to achieve high product quality and low SEC [17].

As the feed becomes more concentrated through the PV, the TDS of the permeate increases from the front to the rear elements. If the permeate of the rear element is blended with the RO feed, the feed is diluted, and a higher-quality permeate can be produced. To minimize the loss of feed pressure, the rear permeate is blended with the SWRO feed before pressurization. For the SSP design, pipelines are required to transport the rear permeate to the SWRO feed. To avoid piping costs, the two PVs are operated side by side, as depicted in Figure 6.12(b). The rear permeate is collected and sent to the SWRO feed tank, where the rear permeate and feed are blended together. The permeate from the front elements is collected and subjected to post-treatment.

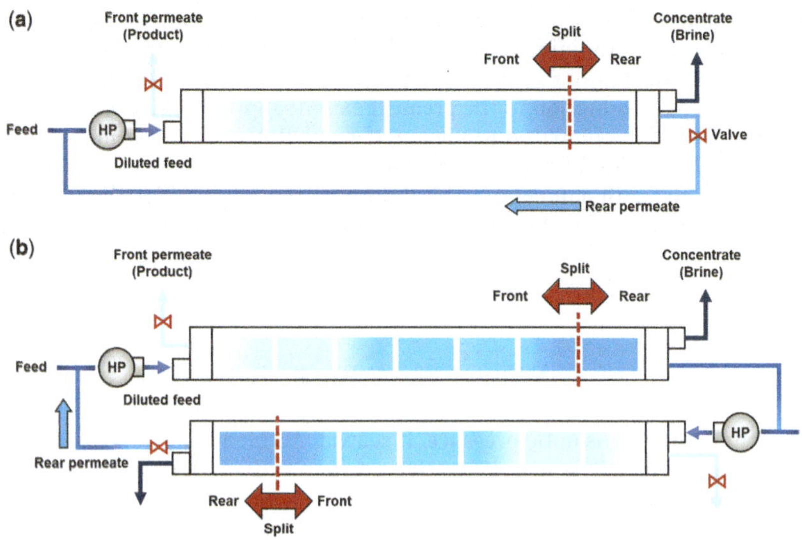

Figure 6.12 Schematic diagram of a split partial single pass (SSP) RO system in SWRO desalination plants: (a) a conceptual SSP RO design and (b) a single unit design to reduce piping costs. This figure was adopted from [17].

The performance of the SSP RO is significantly influenced by the number of split elements because the permeate from the split elements leads to the osmotic dilution of the feed. As shown in Figure 6.13(a), the feed TDS concentration (35 000 mg/L) decreases to 34 375 and 30 000 mg/L at SSP 7 and SSP 4–7 splits, respectively. Therefore, an initial osmotic pressure decreases from 29.68 bar to 29.15 and 25.44 bar, thereby resulting in a reduced hydraulic pressure required for the target water recovery. SSP RO has the potential to consume less energy than single-pass RO. However, the SEC of the SSP RO was higher than that of the single-pass RO because of the loss of rear permeate through circulation.

As depicted in Figure 6.13(b), SSP 7 requires 2.17 kWh/m^3 and other SSP RO systems need higher SEC. The single-pass RO consumes 2.11 kWh/m^3, while the two-pass RO utilizes 2.22–2.71 kWh/m^3. Therefore, SSP 7 is only economically feasible because it is in the range between 2.11 and 2.22 kWh/m^3 even though SSP RO designs can meet the target water quality. The SEC of the RO systems with respect to recovery rates (30–50%) is presented in Figure 6.13(c). Regardless of recovery rates, SSP 7 exhibits higher SEC (2.17–2.29 kWh/m^3) than that (2.10–2.26 kWh/m^3) of single-pass RO, but lower SEC than that of two-pass RO (i.e., 2.20–2.40 kWh/m^3 for SPSP, 2.37–2.51 kWh/m^3 for partial two pass, and 2.71–2.83 kWh/m^3 for two-pass RO). Figure 6.13(c) also indicates that the SEC of single-pass RO becomes more similar (ranging from 0.13 to 0.03 kWh/m^3) to that of SSP RO as the recovery rate increases. The practical application of SSP RO should be further investigated, but SSP RO will be a promising new technology in an era of water scarcity because of its high-water quality and low energy expenditure.

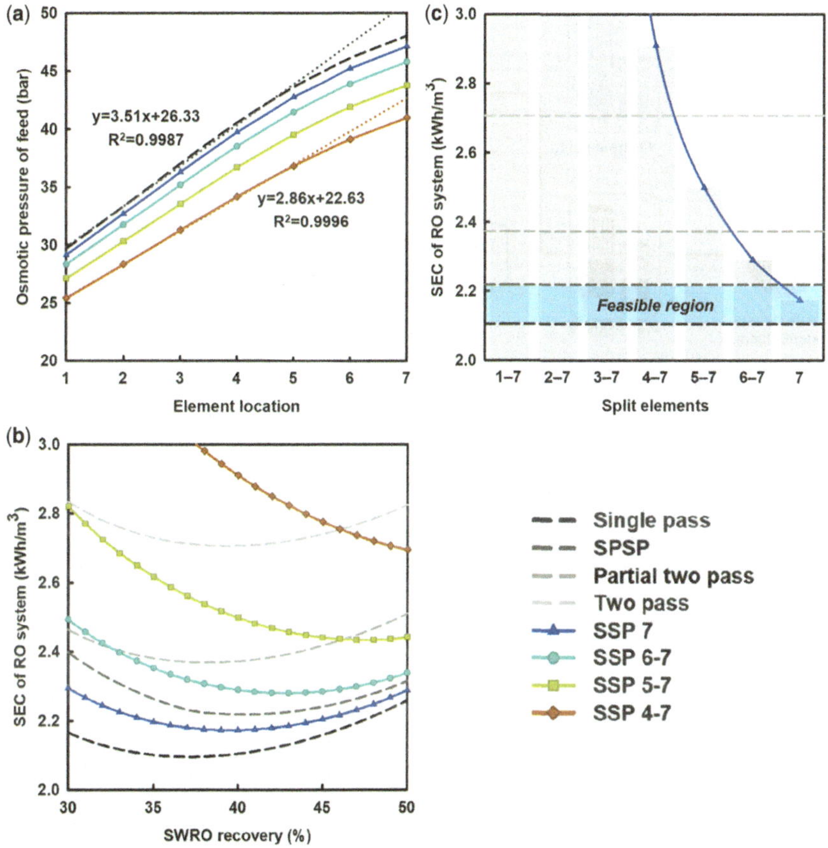

Figure 6.13 (a) Feed osmotic pressure of the elements in a pressure vessel, (b) energy requirement depending on the split ratio in the SSP RO system, and (c) energy requirement of different RO systems under various SWRO recovery rates. This figure was adopted from [17].

6.3.2 FO and reverse osmosis (RO) hybrid process

FO, which is an osmotically driven separation process, utilizes a high-concentration draw solution (DS) and a low-concentration feed solution (FS) for the generation of an osmotic pressure gradient to drive the transport of pure water. Because of the low applied pressure, FO is considered to be a low-energy desalting process [359], but the requirement of an additional recovery process to re-concentrate the diluted DS and produce fresh water results in high energy consumption. Therefore, the FO–RO hybrid process depicted in Figure 6.14 was proposed as a novel energy-efficient hybrid desalting process [360]. In FO, seawater used as the DS is diluted and then supplied to the RO. Simultaneously, low-salinity water (e.g., municipal and industrial wastewater, brackish water, and groundwater) is utilized as an FS to create a driving force (osmotic pressure gradient). Therefore, the FO–RO hybrid process has a significant benefit in

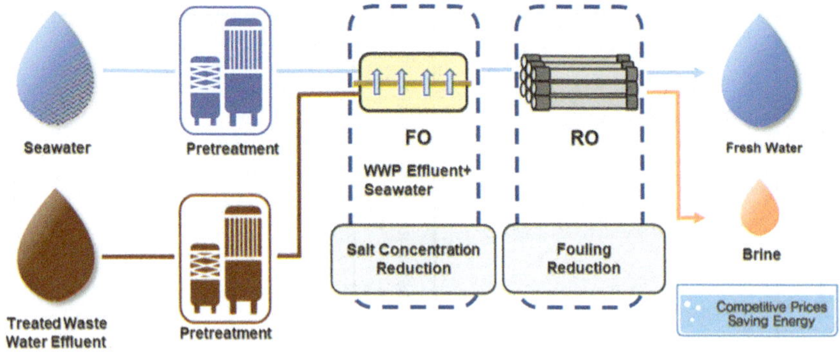

Figure 6.14 Schematic diagrams of the FO–RO hybrid system for integrating wastewater reclamation and seawater desalination.

extracting clean water from the feed using free osmotic pressure, leading to partial desalination (diluted seawater) that can be further desalinated by a subsequent RO step to produce clean water. The concentrated impaired water can be further dewatered to recover nutrients for use as fertilizer or biogas recovery by an anaerobic process. This process is expected to greatly enhance the sustainability of both wastewater reclamation and seawater desalination by reducing energy consumption and environmental impact [361, 362]. As FO acts as a pretreatment of wastewater effluent prior to the RO stage, the fouling potential shifts from RO to FO. Coupled desalination processes may achieve higher system recoveries by operating at higher pressures or temperatures without the risk of scaling.

Pilot-scale seawater desalination plants of the FO–RO hybrid process with wastewater from a coal-fired power plant as FS and the SWRO process were operated for a five-month period [361]. Figure 6.15 depicts the pilot plant operation results (i.e., flow rate, TDS, SEC, and SDI) of both systems with similar desalinated amounts of TDS [i.e., 856 800 (FO-RO) and 857 398 (SWRO) g/day] during summer. For the FO–RO hybrid process, the FO system

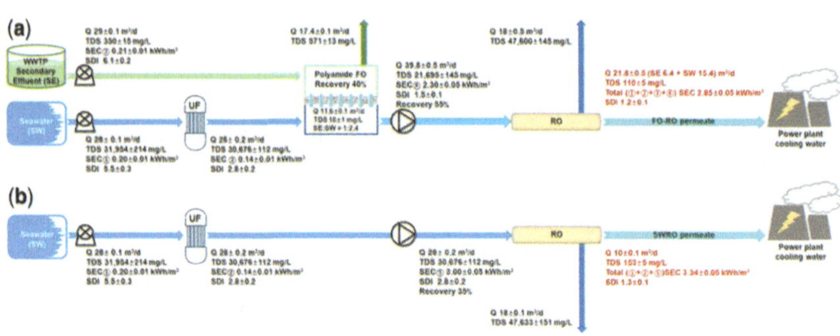

Figure 6.15 Schematic summary showing flow rate, TDS removal, SEC, SDI and recovery in (a) FO–RO hybrid and (b) SWRO. This figure was adopted from [361].

treating wastewater at 40% recovery used $0.55 \pm 0.01\,kWh/m^3$ including a UF seawater pretreatment. The subsequent RO was operated at 55% recovery, while consuming $2.3 \pm 0.05\,kWh/m^3$ of energy. On the other hand, the SWRO system consumed $3.0 \pm 0.05\,kWh/m^3$ at a recovery of 35%. As a result, the SEC for the FO–RO hybrid process ($2.85 \pm 0.05\,kWh/m^3$) was 23.3% less than that for a typical SWRO desalination ($3.34 \pm 0.05\,kWh/m^3$), which implies better economic feasibility in terms of operational energy consumption.

Besides FO, pressure-retarded osmosis (PRO) also has the potential to be combined with RO for the dilution of seawater [363]. In PRO, pure water is transported from low-salinity water (FS) to pressurized high-salinity water (DS). Hence, PRO theoretically plays the role of osmotic dilution and energy production. In particular, the additional energy produced through PRO can power the RO operation and further reduce the energy consumption of the entire PRO–RO hybrid process, compared to the FO–RO hybrid process.

Figure 6.16 indicates the estimated SECs of the SWRO, FO-RO, and PRO-RO systems with or without ERDs. FO–RO and PRO–RO systems without ERDs were 2.1 and $1.6\,kWh/m^3$, respectively, while the SWRO system exhibited $3.9\,kWh/m^3$. With ERDs, the SECs of all systems exhibited the significantly reduced SECs to be 1.4, 1.1, and $2.3\,kWh/m^3$ for the FO–RO, PRO–RO, and SWRO systems, respectively. The moderately low SEC in the FO–RO hybrid process was attributed to the diluted seawater by FO, which enables low hydraulic pressure in RO for the production of equivalent water. On the contrary, in the PRO–RO hybrid process, the conversion of chemical energy to mechanical energy, as well as the dilution of seawater, could further reduce energy consumption.

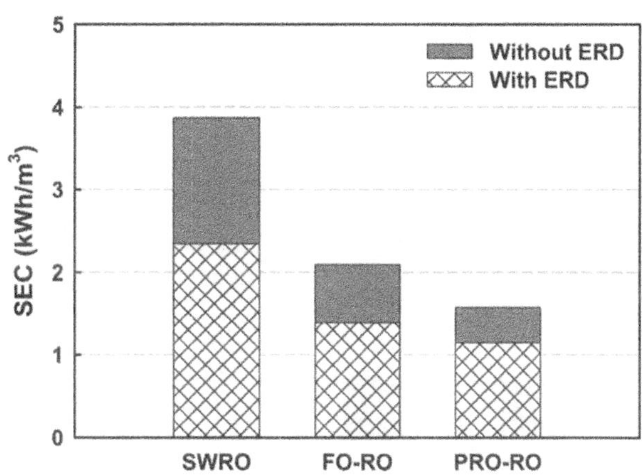

Figure 6.16 SEC evaluation of FO–RO and PRO–RO hybrid processes with and without PXs as ERDs. The estimated SEC results are defined as overall energy consumption (kWh) per water produced (m^3). This figure was adopted from [363].

Despite the high potential of the FO–RO and PRO–RO hybrid processes for the reduction in energy consumption, both processes have many challenges that need to be overcome. The small osmotic pressure difference between the two streams results in a low membrane flux, thereby requiring additional investment costs associated with FO units and a larger membrane area, which makes economic sustainability unquestionable.

6.3.3 Draw solution-assisted RO

For high water recovery and energy efficiency, a concept using the balance of the two different driving forces (i.e., hydraulic pressure and osmotic pressure) was proposed and variously referred to as draw solution-assisted RO (DSARO), osmotic-assisted reverse osmosis (OARO), osmotically enhanced dewatering (OED), and cascading osmotically mediated reverse osmosis (COMRO). DSARO employs the DS to not only induce the transfer of spontaneous water molecules through a high concentration of DS, but also reduces the osmotic pressure difference, as depicted in Figure 6.17(a). As a result, DSARO can reduce the

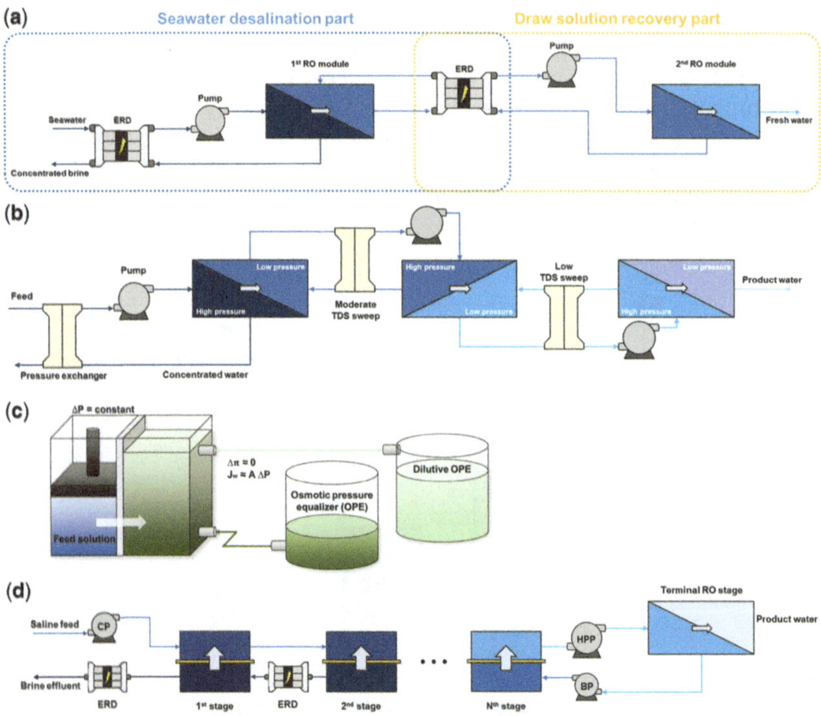

Figure 6.17 The conceptual illustration of (a) draw solution assisted reverse osmosis (DSARO) (This figure was adopted from [33].), (b) osmotically assisted reverse osmosis (OARO) (This figure was adopted from [355].), (c) osmotically enhanced dewatering (OED) (This figure was adopted from [29].), and (d) cascading osmotically mediated reverse osmosis (COMRO) (Reprinted with permission from [22]. Copyright 2018 American Chemical Society.).

high-pressure requirement in the conventional RO process; therefore, a two-stage low-pressure RO configuration is adopted in DSARO. Although the SEC of DSARO is higher than that of the conventional RO process, the results of an economic analysis indicate that the specific water production cost of DSARO can be 10% lower than that of the conventional RO process [33].

Figure 6.18 presents the effect of operation variables (i.e., DS concentration, flow rate ratio, and membrane structural parameter) on SEC in DSARO. An increase in DS concentration leads to a decrease in the applied pressure in the first RO and the increased pressure of the second RO, thereby resulting in reduced SEC in the first RO and increased SEC of the second RO, as shown in Figure 6.18(a). As a result, the overall SEC slightly increases with increasing DS concentration. Flow rate ratios exhibit a more significant impact on SEC in DSARO, as presented in Figure 6.18(b), because they influence the change in DS concentration in the RO modules. In the case of a low flow rate ratio, the DS concentration is dynamically changed at the first RO module, thereby leading to a higher applied pressure for the required first RO recovery. This low DS flow rate might reduce the energy requirement of a high-pressure pump,

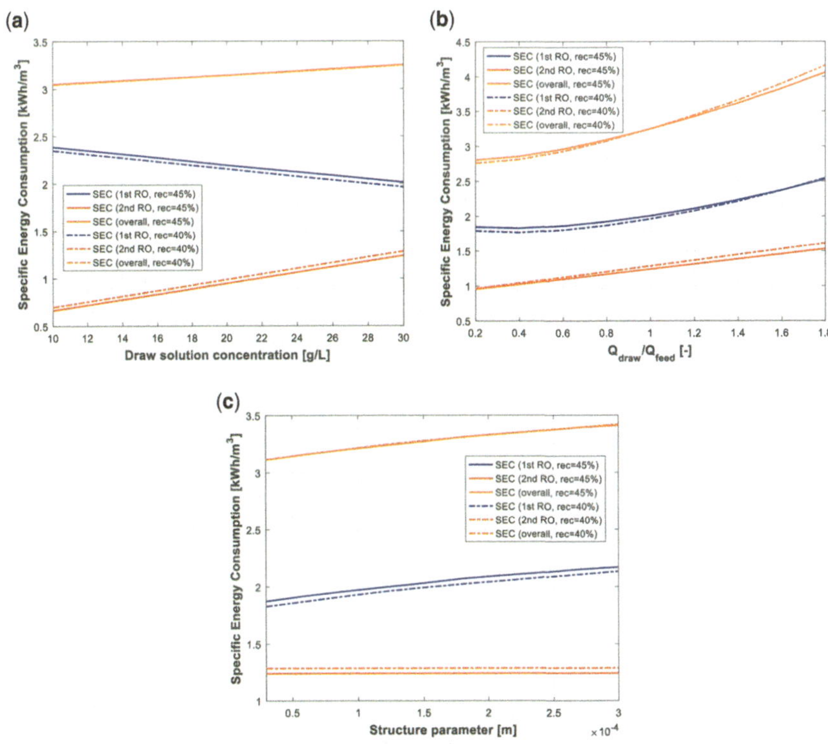

Figure 6.18 Effects of (a) DS concentration, (b) flow rate ratio, and (c) membrane structure parameter on the SEC of each RO stage in DSARO. This figure was adopted from [33].

but a high recovery rate of second RO could increase the applied pressure. The structure parameter of the FO membrane also affects the SEC in DSARO because it changes the internal concentration polarization (ICP), as presented in Figure 6.18(c). When the structure parameter decreases, the ICP is mitigated, thereby reducing the required pressure in the first RO. However, if the structure parameter is too low, the membrane would not withstand a high applied pressure. In addition, membranes with both thin and strong properties do not exist in the current state.

As mentioned above, several configurations similar to those of DSARO have been proposed. In OARO, as shown in Figure 6.17(b), the osmotic pressure difference across the membrane is reduced with a permeate side saline sweep. A series of OARO stages can be used to sequentially reduce the feed concentration until fully desalinated water can be obtained using the conventional RO process. The OARO process was observed to obtain reasonable recoveries for reduced or comparable energy consumption as the MVC process, which is the dominant high-salinity brine treatment technology [355].

In the OED, the DS with a lower concentration than the FS replaces a portion of the hydraulic pressure as the driving force with the osmotic pressure in Figure 6.17(c). Therefore, compared with the traditional RO process, the OED process can be operated at a lower hydraulic pressure, thereby improving energy efficiency and improving the ability to extract water from saline wastewater [29]. According to a recent study, the OED process was found to have a 35–50% water recovery with an energy consumption of 6–19 kWh/m^3 of product water for an FS of 100–140 g/L NaCl.

To overcome the limitations of conventional RO for hypersaline brine desalination and high recovery seawater desalination, COMRO adopts a novel design of bidirectional countercurrent RO stages to depress the hydraulic pressure required by reducing the transmembrane osmotic pressure difference and simultaneously achieving energy conservation, as shown in Figure 6.17(d). Instead of the 137 bar required by conventional RO to desalinate 70 000 mg/L TDS hypersaline feed, the highest operating pressure in COMRO is only 68.3 bar. In addition, COMRO can save up to 17% of energy [22].

6.3.4 Energy comparison of reducing the osmotic pressure of seawater

As shown in Figure 6.2, the salinity of the feed water is one of the most significant factors in determining the theoretical minimum amount of energy. Hence, reducing the osmotic pressure of seawater (i.e., reducing the salinity of the feed water) is critical for achieving lower energy consumption. Table 6.4 presents a comparison among three processes (i.e., SSP RO, FO-RO, and DSARO) proposed for reducing the osmotic pressure of seawater based on the literature.

When comparing the three processes, SSP RO exhibits the lowest SEC, followed by FO–RO and DSARO, while the optimum recovery rate of SSP RO is the lowest among the three processes, as shown in Table 6.4. Compared to the optimal recoveries of SSP RO and FO-RO, it cannot be easily examined that SSP RO has lower SEC than FO–RO. Considering that high recovery

Table 6.4 Comparison among three processes for reducing the osmotic pressure of seawater.

Process	Minimum Energy consumption	Optimum recovery rate	Feed concentration	Ref.
SSP RO	2.1–2.26 kWh/m^3	37%	30 000 mg/L	[17]
FO-RO	2.85 kWh/m^3	40% for FO and 55% for RO	31 954 mg/L	[361]
DSARO	3.25 kWh/m^3	45%	35 000 mg/L	[33]

increases SEC significantly as shown in Figure 6.2, FO–RO may have higher energy efficiency than the other two processes. The main contribution of the high energy efficiency in FO–RO is the effect of dilution in the feed water by the implementation of the FO system. In addition, a variety of parameters, such as the economic feasibility, feed water quality, and product water quality should be evaluated for a fair comparison because of their different applications. Even though SSP RO is the most economical, it should be noted that its optimal recovery rate is the lowest among the three processes.

When comparing FO–RO with DSARO, FO–RO exhibits lower energy consumption than DSARO but requires more membranes and equipment because it is composed of two separate systems. Hence, FO–RO will require a higher capital cost for the construction of two systems and a higher maintenance cost for the replacement of both RO and FO membranes. In terms of the quality of product water, FO–RO shows the highest quality because FO dilutes seawater, thus RO can desalt seawater with reduced salinity. Therefore, the optimum process for reducing the osmotic pressure of seawater should be selected according to the operating conditions and requirements.

Although DSARO process has the highest SEC for seawater desalination, this process can reduce the maximum applied pressure quite significantly, which may also reduce membrane fouling. In addition, DSARO can be operated without any external feed stream, such as for the wastewater in FO–RO. These two features highlight that DSARO has their own advantages for treating hypersaline wastewater or seawater.

It has been reported that FO–RO can show very low SEC approximately 1.3–1.5 kWh/m^3 [364, 365]. However, it should be noted that FO–RO requires an additional low-saline stream which is usually provided from a wastewater or a brackish water stream. Due to the existence of the additional low-saline stream, two problems should be considered for the implementation of FO–RO process. First, FO process in the FO–RO needs a large amount of wastewater for driving the FO process effectively. Second, the implementation and operation of the FO process generates the additional capital and operating costs. In Figure 6.19, the correlation between the ratio of dilution in the feed stream by the FO process and the corresponding theoretical minimum energy reduction is displayed. To reach a ratio of dilution in the feed at 33.33%, the amount of permeate flow in the FO process is half of the feed flow rate. At this condition, the theoretical minimum energy can be reduced from 1.07 to 0.66 kWh/m^3 due

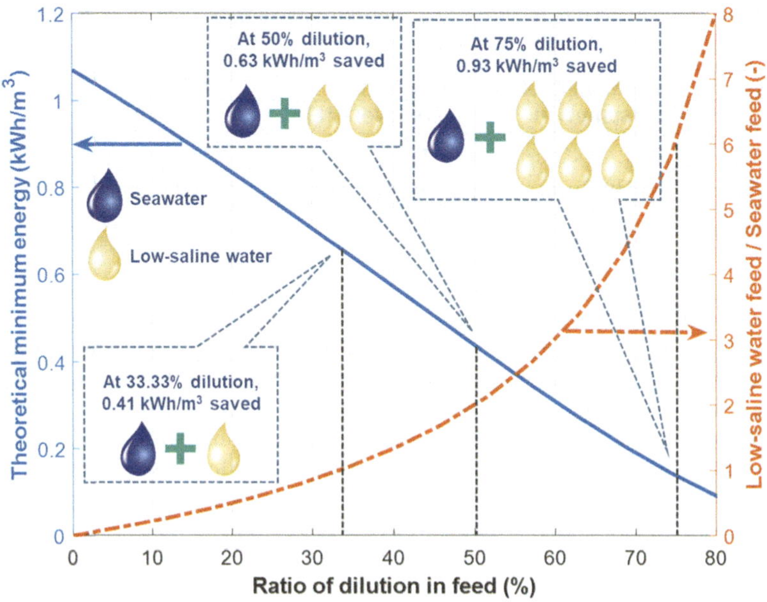

Figure 6.19 Theoretical minimum energy and low-saline water requirement in the FO–RO hybrid according to the ratio of dilution in the seawater feed stream. The FO recovery is assumed to be 50% in this graph. (This figure was adopted from [12].)

to the dilution effect in the feed solution. However, if the FO recovery in the FO–RO could be assumed to be 50% (actually, this is not easily achieved in large-scale operation [366]), the wastewater or low-saline water feed flow rate should be the same as the flow rate of the seawater feed. To increase the ratio of dilution at 50%, the amount of wastewater feed should be approximately twice that of the seawater feed stream, although the theoretical minimum energy can be saved approximately 0.63 kWh/m³. It is very difficult to obtain this amount of wastewater, especially in a large-scale seawater desalination. In addition, additional energy consumptions for driving the FO process such as circulation pump and pretreatment energy in the FO should be considered for calculating the precise SEC of the FO–RO. Some reports have provided the information of the additional energy consumption in the FO. The energy consumption in the FO process and pretreatment of wastewater would be approximately 0.1–0.2 kWh/m³ [16, 313]. Therefore, even though the energy consumption of the FO–RO hybrid process may be the lowest among the reviewed processes to decrease the osmotic pressure of seawater, these limitations provide a large barrier for a wide implementation of the FO–RO for seawater desalination.

In summary, the improvement axis of decreasing the osmotic pressure of seawater is not easily achieved even though the possibility of the improvement axis is the largest among the three suggested axes. Even if the same amount of wastewater as the seawater feed can be fed into the FO–RO, the SEC reduction expected by the dilution of feed solution might be only 0.21–0.31 kWh/m^3. The other methods such as SSP RO and DSARO cannot reduce the SEC of SWRO practically.

6.4 OSMOTIC ENERGY RECOVERY IN CONCENTRATE STREAMS

Salinity gradient power is known as a renewable and sustainable energy source because mixing two streams with different salinities (e.g., freshwater and saltwater) releases the free energy. For example, the potential energy from the mixing of river water and saltwater in the world exceeds 2 TW. This energy is significantly influenced by the salinity gradient between the two streams; therefore, a high-concentration solution has the benefit of producing high energy. Membrane technologies, such as RO, produce highly concentrated brine as well as pure water during seawater desalting. If highly concentrated brine can be utilized to recover osmotic energy, the energy requirement for seawater desalination will be further reduced. For this, two technologies [i.e., pressure-retarded osmosis (PRO) and reverse electrodialysis (RED)] have been proposed to extract this energy by mixing RO brine and low-salinity solutions (e.g., river water, municipal water, and brackish water). This section discusses these technologies for harvesting osmotic energy.

6.4.1 Pressure retarded osmosis

In the 1970s, Sidney Loeb first proposed PRO technology as an energy generation technology utilizing the osmotic pressure gradient [367]. In PRO, the less concentrated stream (FS) and the more concentrated stream (DS) are supplied to two separate flow channels. The difference in the concentration of the two streams induces different chemical potentials and generates an osmotic pressure gradient ($\Delta\pi$). This leads to pure water flow from the low-concentrated FS to the highly concentrated DS via a semi-permeable membrane, thereby resulting in the dilution of the DS and the concentration of the FS simultaneously until the chemical potential of the two streams reaches the equilibrium state [368].

The spontaneous mixing of two streams with the osmotic pressure gradient releases the free energy, which is called the Gibbs free energy (ΔG_{mix}). The Gibbs free energy is the maximum theoretical energy for energy production. The general equation of ΔG_{mix} of the two different salinity solutions in an ideal process that can be expressed as Equation (6.2). When pure water moves from the FS to the DS under hydraulic pressure, the osmotic pressure is converted into hydraulic pressure [368].

Ideal PRO membranes have a water flux, as expressed in Equation (6.32) in terms of the osmotic pressure difference and the applied pressure difference [368]. Salt-rejecting membranes have external concentration polarization (ECP), in which rejected salts accumulate near the membrane surface. In addition, as PRO membranes consist of an active layer and a highly porous support layer, the support layer hampers the transport of salts, inducing internal concentration polarization (ICP). ICP has a significant impact on the osmotic pressure difference across the PRO membranes. Therefore, Equation (6.32) can be modified into Equation (6.33), thereby reflecting both ICP and ECP [369].

$$J_w = A(\Delta\pi - \Delta P) \tag{6.32}$$

$$J_w = A\left\{\frac{\pi_D \exp\left(-\left(J_w S / D\right)\right) - \pi_F \exp\left(\left(J_w / k\right)\right)}{1 + \left(B / J_w\right)\left[\exp\left(\left(J_w / k\right)\right) - \exp\left(-\left(J_w S / D\right)\right)\right]} - \Delta P\right\} \tag{6.33}$$

where J_w is the water flux, A is the water permeability coefficient of PRO membranes, B is the salt permeability coefficient of PRO membranes, $\Delta\pi$ is the osmotic pressure gradient between DS and FS, π_D is the bulk osmotic pressure of the DS, π_F is the bulk osmotic pressure of the FS, ΔP is the applied pressure, D is the diffusivity, k is the mass transfer coefficient, and S is the structural parameter. The theoretical energy produced through PRO can be quantified in terms of the power density and power generated per unit membrane area (W/m^2). The power density is a function of the water flux and applied pressure because chemical potentials are converted into hydraulic energy through water transport. Therefore, the power density can be expressed simply using Equation (6.34) [370].

$$W = J_w \Delta P = A(\Delta\pi - \Delta P)\Delta P \tag{6.34}$$

where W is the power density. Figure 6.20 shows the effect of the applied pressure on the power density in PRO. As the applied pressure increases, the power density is enhanced, but it starts to decrease as ΔP exceeds half of the osmotic pressure. This indicates that the power density can be maximized when the hydraulic pressure is half of the osmotic pressure. In addition, the maximum power density is directly proportional to the water permeability and the square of the osmotic pressure difference across the membrane.

PRO has high potential to be integrated with SWRO (RO-PRO), as presented in Figure 6.21, for energy production, reducing the operational

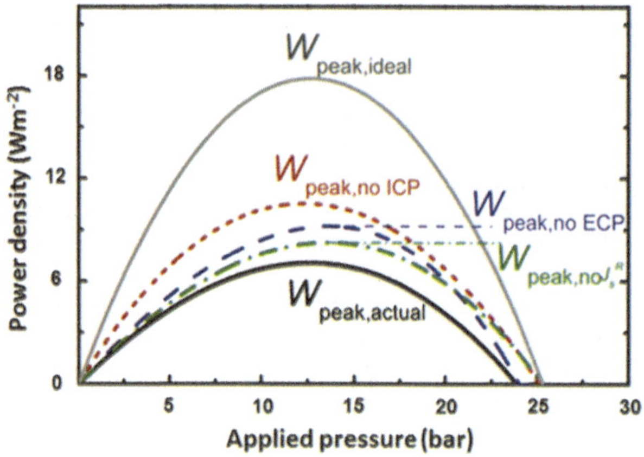

Figure 6.20 Power density (W) during PRO operation as a function of applied hydraulic pressure (ΔP). Reprinted with permission from [125]. Copyright 2011 American Chemical Society.

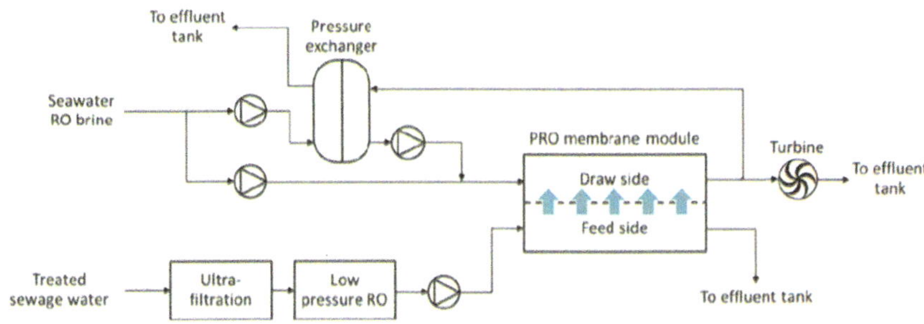

Figure 6.21 Conceptual diagram of the RO–PRO hybrid system of the Mega-ton project in Japan. This figure was adopted from [371].

energy requirement and minimizing the environmental impact of direct ocean discharge [363]. In PRO, a highly saline RO concentrate has high osmotic pressure, and this energy can be harvested using ERD. Therefore, RO–PRO can reduce energy expenditure and operating costs, and two RO–PRO strategies have been proposed. The first is the generation of power through a hydro turbine. This method exhibits low electricity production efficiency owing to the utilization of the hydro turbine; however, the electricity produced from this system can be used for other purposes. The second is the circulation of energy through a pressure exchanger. This system is highly efficient owing to the lack of energy conversion, but the energy or pressure recovered from this system can only be used in the operation of SWRO.

Four different configurations of the hybrid RO–PRO were proposed by Kim *et al.* [372]. Among them, two configurations are RO–PRO hybrid systems for power generation, and the others are for freshwater production. The two configurations for osmotic energy recovery in the concentrate stream are shown in Figure 6.22. The first configuration (Figure 6.22a) utilizes RO to

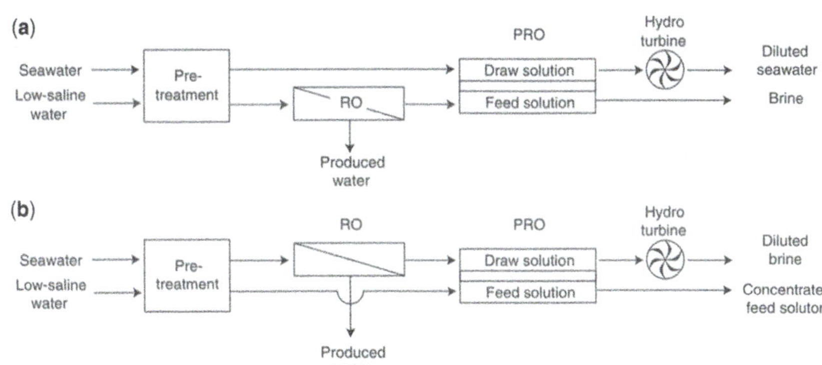

Figure 6.22 Schematic of the proposed RO–PRO hybrid system configurations: (a) low-saline water as an RO feed and (b) seawater as an RO feed. This figure was adopted from [372].

treat low-saline water, such as municipal wastewater, and then the RO brine is supplied to the feed channel. Simultaneously, seawater is used as the DS. This configuration yields relatively lower energy compared to the other configurations because the driving force in PRO is significantly reduced, owing to the highly concentrated RO brine. On the other hand, the second configuration (Figure 6.22b) produces water from seawater, and then brine is used as the DS with low-saline water as the FS. Therefore, this configuration has the highest power generation because the concentrated RO brine with a high concentration can effectively transport water from the FS to the DS.

Achilli *et al.* evaluated a pilot-scale RO–PRO system consisting of an SWRO system, a PRO system, an ERD, and a pressure exchanger [373]. Seawater pressurized by a pressure exchanger is supplied to the RO modules, and the concentrated seawater brine generated from the RO system reaches the ERD. In PRO, seawater brine and wastewater effluent are utilized as the DS and FS, respectively. Finally, the diluted RO brine transfers its energy into the seawater RO feed via the pressure exchanger and is then discharged or reused. This configuration could achieve higher power density (1.1–2.3 W/m^2) than the river-to-sea PRO pilot systems (1.5 W/m^2).

In Japan, the prototype integrated PRO hybrid process was evaluated through a Mega ton water project, as shown in Figure 6.21 [371]. This hybrid system could achieve a power density of 13.8 W/m^2 at an operation pressure of 30 bar. The feed water discharge was applied to provide circulation for the feed and mitigate solute leakage and ICP. During operation, it was observed that the maximum power density was 13.3 W/m^2 and the energy savings were 10–30%.

In South Korea, the GMVP project was launched to evaluate the large-scale RO–PRO hybrid system (240 m^3/d) using wastewater treatment plant effluent and SWRO brine as the FS and DS, respectively, for a low-cost seawater desalination operation [374, 375]. Instead of focusing on only the energy production, the GMVP examined not only the energy production, but also the recovery of the pressure from the pressurized DS. In the SWRO–PRO hybrid system, the osmotic power was recovered using a hydraulic turbine and ERD. The maximum power density for this implementation was 18.3 W/m^2, and the energy expenditure was reduced to 80%.

A PRO, RO, and NF integrated system was also proposed, as shown in Figure 6.23 [376]. Wastewater effluent was first treated using NF, and then the permeate was supplied as the FS to PRO. Seawater was fed into the SWRO and the concentrated RO brine was used as the DS. Therefore, the overall energy consumption of the hybrid system was reduced by approximately 15%. In addition, the permeate can be applied for irrigation, thus, this system has the potential for combined wastewater treatment, seawater desalination, and energy production for the food–water–energy nexus.

6.4.2 Reverse electrodialysis

Pattle first proposed the concept of RED for the production of electric power in 1954 [377]. Figure 6.24 illustrates a standard RED unit with a membrane stack and electrodes. The membrane stack consists of alternately stacked cation exchange membranes (CEMs) and anion exchange membranes (AEMs),

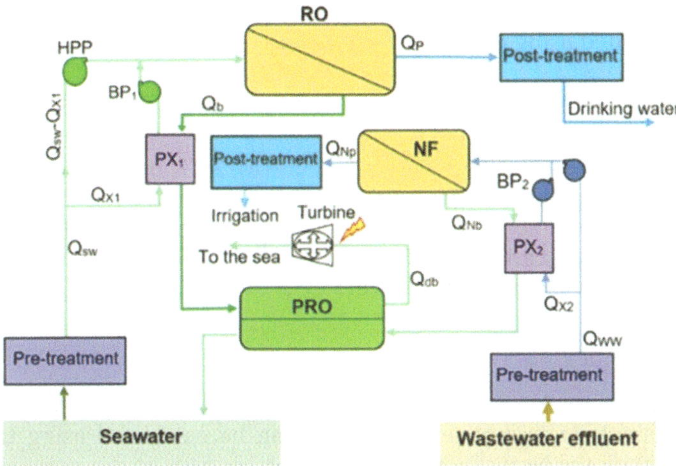

Figure 6.23 The RO–NF–PRO hybrid process for wastewater treatment, seawater desalination, and energy production. This figure was adopted from [376].

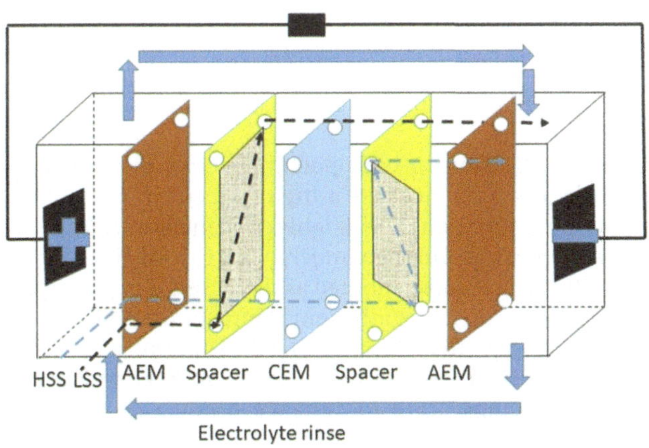

Figure 6.24 Schematic diagram of a RED module. This figure was adopted from [379].

and high-salinity and low-salinity compartments are formed for a high-salinity solution (HSS) and a low-salinity solution (LSS), respectively. AEMs are selective for anions, whereas CEMs allow the transport of cations. The difference in concentration between the HSS and LSS induces the selective transport of ions from the HSS towards the LSS through CEMs and AEMs. Cations are transported towards the CEM side and anions toward the AEM side, creating an ionic flux to be converted at the electrodes by using a redox reaction to generate electricity [378].

In the RED, the power density (P) dissipated by the external resistance R_L can be calculated using Equation (6.35) [380].

$$P = I^2 R_L = \left(\frac{E_{emf}}{R_{stack} + R_L}\right)^2 R_L \tag{6.35}$$

where I is the current in the electrical circuit (A), R_L is the external resistance (Ω), E_{emf} is the theoretical electromotive force (V), and R_{stack} is the sum of the ohmic resistance of stack components and non-ohmic resistance. The maximum power density (P_{max}) can be obtained using Equation (6.36) when the resistance of the external load equals the internal stack resistance ($R_L = R_{stack}$).

$$P_{max} = \frac{E_{emf}^2}{4 N_m S_m R_{stack}} \tag{6.36}$$

where N_m is the number of IEMs, and S_m is the projected area of the IEMs (m^2). The theoretical electromotive force (E_{emf}) can be calculated using Equation (6.37), based on the Nernst equation [378].

$$E_{emf} = \frac{N_m \alpha RT}{zF} \ln\left\{\frac{\gamma_{HSS} C_{HSS}}{\gamma_{LSS} C_{LSS}}\right\} \tag{6.37}$$

where α is the average membrane permselectivity, F is the Faraday constant (96 485 C/mol), R is the universal gas constant (8.314 J/mol K), T is the absolute temperature, C_{HSS} and C_{LSS} are the molar concentrations of HSS and LSS, respectively, γ_{HSS} and γ_{LSS} are the activity coefficients, and z is the ionic valence. Other parameters for the calculation can be obtained from the literature [378, 381, 382].

Similar to PRO, RED has a high potential to be combined with SWRO, as illustrated in Figure 6.25, because a high concentration of the RO brine is beneficial for energy production [379]. Highly concentrated RO brine is generated during SWRO desalination and then undergoes RED for electricity production. In RED, the SWRO brine can be adopted as the HSS for high-power production as well as a reduction in the negative environmental impact of SWRO brine discharge. Electricity from the RED is utilized to operate electrical equipment, such as pumps, in SWRO. In addition, heating by solar energy can be employed to increase the feed temperature to enhance production and lower the feed pumping energy requirements. High temperatures can significantly enhance the efficiency of RED owing to an increase in osmotic pressure and activity coefficients and a decrease in membrane resistance and solution viscosity.

Because a simple combination of a RED unit with an RO unit can be limited to realizing the commercial application of RED–RO, a complex configuration of RED–RO is required for practical applications [379]. Four configurations of RED–RO were proposed by Li *et al.*, as depicted in Figure 6.25 [383]. The first configuration (Figure 6.25a) proposed that seawater and secondary effluent from wastewater treatment plants are supplied as the HSS and LSS, respectively, for RED, and then diluted seawater is treated by RO. This configuration can be modified as presented in Figure 6.25(b) to recycle the RO brine into seawater and then increase the salinity of seawater for enhanced power production in RED. Figure 6.25(c) shows the osmotic energy recovered from the RO brine.

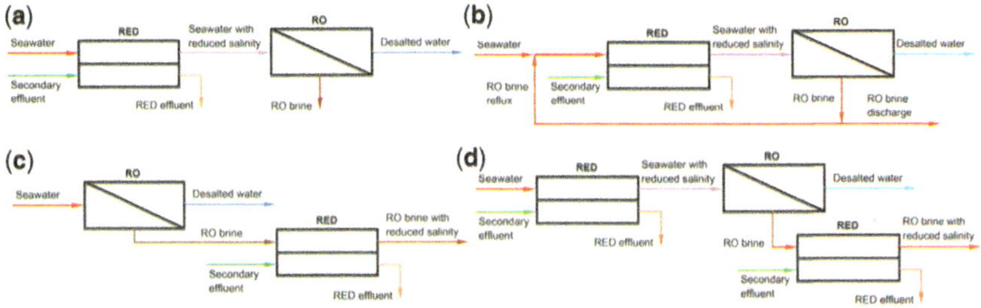

Figure 6.25 Schematic diagram of four different SWRO–RED hybrid configurations. This figure was adopted from [383].

Highly concentrated RO brine is employed as the HSS, and secondary effluent is used as the LSS, which generates electricity. Simultaneously, the RO brine is diluted, and the environmental impact of the discharge is mitigated. Both configurations from Figures 6.25(a) and 6.25(c) can be integrated as depicted in Figure 6.25(d) to maximize the benefit of RED for SWRO desalination. The results of this study indicate that all RED–RO configurations are superior to conventional SWRO because all configurations can reduce the total energy consumption and realize a zero-discharge system with higher recovery.

An interesting application of RED–RO was proposed for energy self-sufficient agriculture in coastal arid regions, as shown in Figure 6.26 [384]. Groundwater is first evaporated and concentrated on the air inlet using an evaporator to maintain the indoor air temperature and high humidity to reduce irrigation water consumption. The condenser then restores a portion of the water vapor as irrigation water on the air discharge outlet. In RED, seawater and highly concentrated groundwater are fed as the LSS and HSS, respectively, for a power supply to RO and other electrical equipment. Seawater is further desalinated through RO, and fresh water can be used as irrigation water.

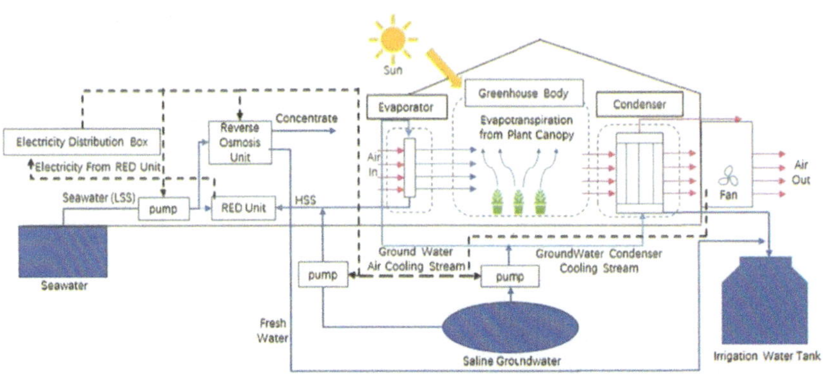

Figure 6.26 Schematic illustration of the sustainable greenhouse system combined with the RO–RED hybrid system. This figure was adopted from [384].

The RO–RED hybrid system can also be integrated with further desalting technologies such as membrane distillation (MD) and capacitive deionization (CDI) to enhance power production. MD can produce high-quality distillate under relatively low operational temperatures, and its performance is less influenced by salinity than that of RO. Therefore, seawater can be further concentrated by an RO–MD hybrid system and provide a higher salinity solution to RED for more energy production [385]. This configuration achieved a maximum exergy efficiency of 49% under the best scenario (MD feed temperature of 60°C, MD brine concentration of 5 M NaCl, RED power density of 1.1 W/m^2), which enabled near-zero liquid discharge and low-energy seawater desalination [386].

CDI desalting technology can also be coupled with SWRO desalination for ultrapure water production [387]. For desalination, CDI utilizes ion electrosorption on the surface of a couple of charged electrodes with an AEM and CEM pair to significantly enhance the charge efficiency and desalting performance by preventing the transport of co-ions. CDI has many advantages such as low energy consumption, small environmental impact, and high water recovery; however, its performance can be restricted when treating high-TDS solutions such as seawater [388]. Therefore, CDI is applied for desalting low-salinity RO permeate to produce ultrapure water. Then, the RO brine and ultrapure water from CDI are supplied to RED for higher electricity generation. The results of this study indicate that the energy consumption of the RO–MCDI–RED unit was reduced by 39.0% and 16.8%, respectively, compared to that of a typical two-pass RO system without and with RED, respectively [387].

6.4.3 Osmotic energy recovery in a concentrate stream

Theoretically, the maximum osmotic energy that can be recovered by PRO and RED is identical because it is determined by the Gibbs free energy when mixing two streams with different salinities. However, the different efficiencies of the conversion from osmotic energy to power led to higher power density in PRO than in RED [389]. As shown in Figure 6.27, the power densities of PRO are 3.1–3.4, 8–10, and 22–32 times higher than those of RED for natural, anthropogenic, and engineered salinity gradients, respectively. This implies that PRO can employ a remarkably smaller membrane area for power

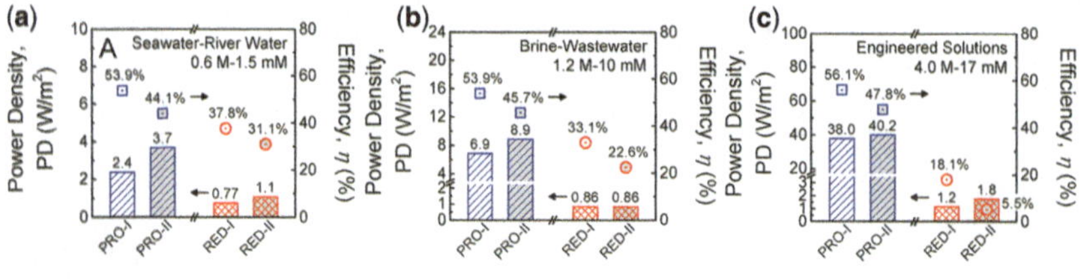

Figure 6.27 Power density (columns, left vertical axis) and efficiency of work extraction (symbols, right vertical axis) for PRO and RED (blue and red data representations, respectively) at 0.6, 1.2, and 4.0 M NaCl. Reprinted with permission from [389]. Copyright 2014 American Chemical Society.

production. Furthermore, the power density of RED is much lower than that of PRO at higher salinity gradients. This high power density performance of PRO is attributed to the inherent characteristics of PRO to effectively utilize larger salinity differences. In contrast, RED cannot obtain high power density benefits from an increased salinity gradient regardless of membrane transport properties, because the Donnan exclusion effect becomes negligible at high solution concentrations, which induces low efficiencies [389]. Consequently, PRO can achieve higher efficiency and power density for a range of salinity gradients compared to RED with existing technologically available membranes. The feasible application of RED energy production is restricted to relatively low salinity gradients. However, despite the high potential of PRO to achieve excellent performance with large salinity gradients, the development of innovative membranes is required for the realization of this technology. In addition, actual power generation is further influenced by the engineering components in PRO and RED. For instance, PRO requires pumps, pressure exchangers, and hydroturbines, whereas RED utilizes a reversible redox couple at the end electrodes and pumping systems. Membrane fouling caused by natural and anthropogenic streams can also detrimentally reduce the efficiency of PRO and RED for power production [390, 391].

The power production and energy consumption of RO–PRO and RO–RED are further compared based on the literature discussed above. The maximum power density for RO–PRO was 18.3 W/m^2, and the energy consumption for the operation of RO–PRO was reduced to 80% [375]. In the case of RO–MD–PRO, the power density of the RED was 1.1 W/m^2, and the maximum exergy efficiency was 49% under an MD feed temperature of 60°C and MD brine concentration of 5 M NaCl. RO–MCDI–RED achieved a 16% reduction in energy consumption compared to the conventional two-stage RO system [387]. From this comparison, RO–PRO exhibits better efficiency in terms of higher power density and lower energy consumption for seawater desalination because of the beneficial properties of PRO for high salinity gradients. However, because the operating conditions of RO–PRO and RO–RED as well as the advantages of PRO and RED are different, the optimum system for osmotic energy recovery from the concentrate stream should be determined based on the environmental conditions (e.g., seawater quality, low-salinity water quality, and capital expenditures).

6.5 IMPROVEMENT OF RO MEMBRANES

The RO membrane is an essential part of the SWRO plant and plays a significant role in the efficient operation of the process. The performance of an RO membrane determines a considerable portion of the operational costs, water production, energy consumption, and permeate quality. Despite the existence of RO membranes for over 60 years, some critical challenges still need to be overcome. These challenges include low removal efficiency against pollutants, energy requirements, and membrane fouling [392]. As these challenges can be linked to the properties of the RO membrane, it is vital to have RO membranes capable of high water flux, high rejection, and resistance to fouling and chemical attack with a minimum energy requirement. To improve the performance of RO membranes, research efforts have focused on chemical replacement during membrane synthesis, membrane surface modification, and the use of new membrane materials.

6.5.1 Introduction of novel building blocks

The use of chemicals such as chlorine is an integral part of the SWRO plant for both pretreatment and post-treatment purposes. Currently used commercial membranes, such as polyamide membranes, are susceptible to chlorine attacks. Over the years, research has focused on making the polyamide membrane chlorine resistant. One way to make polyamide membranes not susceptible to chlorine attacks is to replace conventional membrane building blocks such as *m*-phenylenediamine (MPD) and trimesoyl chloride (TMC), which are commonly used in the fabrication of thin-film composite membranes. A study by Hong *et al.* [393] reported an amide linkage replacement during TFC RO membrane synthesis with the use of imide linkage to enhance the chlorine resistance of RO membranes. In another study, MDP and TMC were completely replaced by *m*-phenylenediamine-4-methyl (MMPD) and cyclohexane-1,3,5-tricarbonyl chloride (HTC) monomers for both chlorine resistance and enhanced membrane performance [394]. Other efforts involve the addition of new additives and the use of hydrophilic amine monomers [395].

6.5.2 RO membrane surface modification

Modifying the surface of the RO membrane is probably the most convenient method for improving the performance of RO membranes without any significant changes in membrane synthesis. Surface modification can be performed physically (adsorption and coating) or chemically (chemical coupling, sol–gel coating, oxidation, hydrolysis, grafting, plasma treatment, etc.) to increase hydrophilicity, surface charge, and surface roughness. The increased hydrophilicity of RO membranes has been shown to increase the water flux and fouling resistance. One way to increase the hydrophilicity of RO membranes by surface modification is by soaking them in various hydrophilic organic solutions such as alcohols and glycerol (physical method) to increase membrane performance, such as water flux, salt rejection, and fouling resistance [396]. However, soaking only temporarily improves membrane performance as the hydrophilic materials attach via weak molecular forces, thus detaching over time. Consequently, the performance decreases over time. This problem can be solved by chemical surface modification, which causes a stronger bond between the RO active layer and the attached materials. Examples of chemical surface modification include partial hydrolysis of the active layer, sol–gel coating, chemical coupling, or grafting [397]. Membrane surface roughness is another membrane property that affects membrane performance. It is generally believed that decreasing membrane roughness is an important strategy for reducing RO membrane fouling propensity [398]. However, surface modification by surface patterning which increases surface roughness has shown a contrary result where making patterns on the membrane is an effective way of improving RO membrane performance [399].

Recently, membrane surface patterning, another surface modification approach, has been explored to enhance the performance of RO membranes. This is because surface patterning can increase the surface-active area, which in turn affects the membrane surface fluid shear. Similarly, hydrodynamic conditions created by some patterns on RO membranes are often unfavorable for microbial colonization, making the pattern membrane an anti-biofouling membrane [399]. Surface patterning of RO membranes has been achieved by template-based micro-molding or direct printing (ink-jet printing and 3D printing) on RO membranes [400].

6.5.3 Biomimetic RO membranes

Recently, extensive research on biomimetic membranes inspired by organisms, which depend on their complex transmembrane proteins for selective permeation of ions and water via channel pores and pumps, depending on the nutrients and energy needs of such organisms, has been shown to have the ability to overcome the permeability-selectivity trade-off of conventional RO membranes, especially polymer-based RO membranes. The potential of these proteins, especially aquaporin, has been used to fabricate future desalination membranes, and promising high-water flux and salt rejection have been reported [401]. Zhao *et al.* [402] prepared an aquaporin-based biomimetic membrane using an interfacial polymerization reaction. The fabricated membrane showed better performance than the BW30 and SW30HR commercial RO membranes.

Although biomimetic membranes using biological organism proteins have not yet been commercialized because of scale-up, mechanical, and cost issues, such membranes, if optimized, still have the potential to achieve better salt rejection and comparable water flux compared to conventional commercialized RO membranes. However, interesting and optimistic results in biomimetic RO membranes have been reported by directly mimicking biological membranes (with respect to complexity, function, and form) by simulating how a biological membrane responds when its surrounding conditions change in terms of the intake and output of ionic species. This has been achieved with the development of membranes that can exhibit distinctive properties in the presence of external stimuli (e.g., chemical, electrical, and physical). These types of membranes are called 'smart' or 'stimulus-responsive' membranes. Smart membranes can modify their functionality to improve their separation and properties, such as water movement, molecular or solute sieving, antifouling, and self-cleaning. Stimuli-responsive agents are polymers [403] or inorganic materials (electrically conductive) [404]. Smart membranes are fabricated using three methods. (1) Integrating a stimulus-responsive polymer within a matrix of the membrane, (2) attaching/coating a stimulus-responsive polymer or material to the surface of the membrane, and (3) using a stimulus-responsive material as a separation barrier. A major drawback of blending stimuli-responsive materials within the membrane matrix is the limited stimulus response. This is because stimuli-response materials are scattered everywhere within the membrane matrix, which loses some of its ability to respond to external stimuli because other membrane materials that are non-responsive are receiving most of the stimuli. However, when the membrane surface is coated or the selective layer is made of materials responsive to stimuli, there is a prompt response to the stimulus.

6.5.4 Inorganic RO membranes

6.5.4.1 One-dimensional and two-dimensional inorganic membranes

To improve the performance of RO membranes, recent research efforts have focused on the development of new RO membranes that would be able to reject solutes by molecular sieving compared to conventional RO membranes that remove solutes using a solution–diffusion-based approach. This new membrane has pores that are usually smaller than the diameter of hydrated ions and subsequently allow only the water molecules to pass through. To achieve this type of molecular sieve, inorganic one-dimensional (1D) and two-dimensional (2D) nanomaterials have been used. These inorganic materials include carbon nanotubes (CNTs), boron nitride nanotubes (BNNTs), graphene, graphene oxides, MXene, transition metal dichalcogenides (TMDCs), covalent organic frameworks (COFs), zeolites, and molybdenum disulfide (MoS_2).

Inorganic membranes are resistant to chlorine and chemical cleaning attacks. Thus, inorganic membranes are a promising alternative to conventional organic membranes (polymeric membranes). The resistance of inorganic membranes to disinfectants and chemicals makes them cleanable with different cleaning agents without any detrimental effects on the membrane. Some inorganic membranes have excellent intrinsic antimicrobial and mechanical properties. Inorganic membranes have thus shown potential for superior water flux, salt rejection, cleanability, antifouling, and mechanical properties (ref). The major challenges currently faced by this new membrane are cost and safety issues. Nanomaterials are quite expensive, and the fate of nanomaterials in the environment is not fully understood. However, the current increased interest in these inorganic nanomaterials in electronics, sensors, energy, and water treatment applications is projected to reduce the cost of these materials owing to increasing demand.

1D nanomaterials, especially CNTs, have been proposed as new membrane materials for seawater desalination. CNTs are vertically aligned to allow water molecules to pass through their tubes, while solutes are rejected. CNTs exhibit ultra-fast water molecules passing along the tubes owing to the atomic-scale smoothness of the nanotube walls and molecular ordering within the nanochannel [405, 406]. Correspondingly, 2D inorganic nanomaterials have been explored for desalination membrane fabrication owing to their easy fabrication and controllable pore size. These membranes are either atomically thin nanoporous membranes or laminar membranes [407]. Atomically thin nanoporous membranes are fabricated by making uniform nanopores on the 2D nanosheets so that solutes are rejected while water molecules can permeate. Laminar 2D membranes are fabricated by the in situ stacking of nanosheets. Nanochannels created at the end of each nanosheet are used for both solute rejection and water permeation.

6.5.4.2 Ceramic membranes

Ceramic membranes for seawater desalination are fabricated from silica, zeolites, alumina, titania, or a composite of these materials. Ceramic desalination membranes, unlike polymeric RO membranes, can withstand extreme temperatures, high concentrations of foulants, and reactive conditions [408]. Although ceramic membranes have been commercialized for microfiltration (MF) and ultrafiltration (UF), the development of ceramic membranes for RO seawater desalination is still ongoing with promising results [409, 410]. The development of ceramic membranes for seawater desalination is still challenging because of its lower performance and higher cost compared to polymeric RO membranes. However, the preparation of composite ceramic RO membranes can improve their performance for membrane-based seawater desalination [411].

6.5.5 Mixed matrix RO membranes

The earlier development of mixed matrix membranes (MMMs) involved gas separation and pervaporation [412]. MMM has been used in seawater desalination over the past years to improve the performance of current polymeric RO membranes [413]. MMM is prepared by blending inorganic material and nanoparticles (e.g., titanium oxide (TiO_2), zeolite, CNT, graphene-based materials, and metals and their oxides) within the membrane matrix. In the case of mixed matrix thin film nanocomposite membranes, either the active layer, the support layer, or both can be blended with nanoparticles or inorganic materials. The incorporation of these materials inside the membrane matrix, especially polymeric membranes, confers the properties of the inorganic (nano) materials on the membrane, such as packing density, permselectivity, mechanical, thermal, chemical stability, and antibiological properties [414]. Significant enhancement of membrane antifouling properties has been reported by adding metals and their oxides inside the membrane matrix. The size, morphology, and type of material incorporated into the membrane matrix determine the type of properties conferred on the membrane.

doi: 10.2166/9781789061215_0131

Chapter 7

Concluding remarks and epilogue

SWRO has extended the possibility of low-energy seawater desalination to a higher level of implementation. The RO technology development in recent decades has led to a tremendous increase in desalination capacity worldwide. The development of RO technology was conducted in various fields of engineering, such as RO membrane development, system configuration optimization, high-efficient equipment design, operation optimization, and improvements of pre- and post-treatment. In this book, we tried to summarize most of the RO developments, especially in terms of energy efficiency. The detailed configuration of conventional SWRO plants and the equipment correlating with energy consumption were introduced. The basic theoretical background of energy consumption in the SWRO system was provided, and the recent trend in the SEC of SWRO plants have been investigated by analyzing the current data on SWRO plants. The main factors affecting the SEC of SWRO plants have been investigated and summarized. To reduce the SEC of SWRO plants further, future directions for low-energy SWRO systems were characterized. The advanced technologies of RO systems which were designed to achieve low-energy desalination have been introduced. Finally, the recent development of RO membranes was summarized.

Currently, desalination communities face great challenges to resolve carbon neutralization and provide a sustainable water supply. SWRO has proposed a valuable solution to these problems by reducing energy consumption and expanding the desalination plant capacity. However, the regulations to achieve carbon neutralization and provide sustainable water supply are getting stronger, the desalination system using SWRO should be developed further to provide a higher level of solutions, such as higher recovery for sustainable water supply and lower SEC for carbon neutralization. Through this book, we provide comprehensive information, knowledge, and analyses about SWRO systems from a basic understanding of RO systems to future directions for the realization of low-energy SWRO systems. In addition, the SWRO plant data included in the graphs show a clear trend of SEC improvement. We believe that the information in this book is helpful for students who want to study desalination systems using RO from basic knowledge, and for engineers who want to overview the current trends of SWRO systems. We hope that this book will serve as a small stepping stone for the SWRO system to improve its current level of technology.

ACKNOWLEDGMENTS

This study was supported by Korea Ministry of Environment (MOE) through Human Resource Development Program for Carbon Neutrality.

doi: 10.2166/9781789061215_0133

References

[1] S. Liu, G. Zhang, M. Han, X. Wu, Y. Li, K. Chen, J. Meng, L. Shao, W. Wei, G. Chen, Freshwater costs of seawater desalination: systems process analysis for the case plant in China, Journal of Cleaner Production, 212 (2019) 677–686, https://doi.org10.1016/j.jclepro.2018.12.012

[2] V.G. Gude, Desalination and water reuse to address global water scarcity, Reviews in Environmental Science and Bio/Technology, 16 (2017) 591–609, https://doi.org/10.1007/s11157-017-9449-7

[3] V.G. Gude, N. Nirmalakhandan, S. Deng, Renewable and sustainable approaches for desalination, Renewable and Sustainable Energy Reviews, 14 (2010) 2641–2654, https://doi.org/10.1016/j.rser.2010.06.008

[4] S. Homaeigohar, M. Elbahri, Graphene membranes for water desalination, NPG Asia Materials, 9 (2017) e427–e427, https://doi.org/10.1038/am.2017.135

[5] N.C. Darre, G.S. Toor, Desalination of water: a review, Current Pollution Reports, 4 (2018) 104–111, https://doi.org/10.1007/s40726-018-0085-9

[6] Y. You, V. Sahajwalla, M. Yoshimura, R.K. Joshi, Graphene and graphene oxide for desalination, Nanoscale, 8 (2016) 117–119, https://doi.org/10.1039/C5NR06154G

[7] K. Reddy, N. Ghaffour, Overview of the cost of desalinated water and costing methodologies, Desalination, 205 (2007) 340–353, https://doi.org/10.1016/j.desal.2006.03.558

[8] S.S. Shenvi, A.M. Isloor, A. Ismail, A review on RO membrane technology: developments and challenges, Desalination, 368 (2015) 10–26, https://doi.org/10.1016/j.desal.2014.12.042

[9] N. Ghaffour, T.M. Missimer, G.L. Amy, Technical review and evaluation of the economics of water desalination: current and future challenges for better water supply sustainability, Desalination, 309 (2013) 197–207, https://doi.org/10.1016/j.desal.2012.10.015

[10] C. Skuse, A. Gallego-Schmid, A. Azapagic, P. Gorgojo, Can emerging membrane-based desalination technologies replace reverse osmosis?, Desalination, 500 (2020) 114844, https://doi.org/10.1016/j.desal.2020.114844

[11] J. Kim, K. Park, D.R. Yang, S. Hong, A comprehensive review of energy consumption of seawater reverse osmosis desalination plants, Applied Energy, 254 (2019) 113652, https://doi.org/10.1016/j.apenergy.2019.113652

[12] K. Park, J. Kim, D.R. Yang, S. Hong, Towards a low-energy seawater reverse osmosis desalination plant: a review and theoretical analysis for future directions, Journal of Membrane Science, 595 (2020) 117607, https://doi.org/10.1016/j.memsci.2019.117607

[13] L.F. Greenlee, D.F. Lawler, B.D. Freeman, B. Marrot, P. Moulin, Reverse osmosis desalination: water sources, technology, and today's challenges, Water Research, 43 (2009) 2317–2348, https://doi.org/10.1016/j.watres.2009.03.010

[14] C. Fritzmann, J. Löwenberg, T. Wintgens, T. Melin, State-of-the-art of reverse osmosis desalination, Desalination, 216 (2007) 1–76, https://doi.org/10.1016/j.desal.2006.12.009

[15] S. Miller, H. Shemer, R. Semiat, Energy and environmental issues in desalination, Desalination, 366 (2015) 2–8, https://doi.org/10.1016/j.desal.2014.11.034

[16] K. Park, H. Heo, D.Y. Kim, D.R. Yang, Feasibility study of a forward osmosis/crystallization/reverse osmosis hybrid process with high-temperature operation: modeling, experiments, and energy consumption, Journal of Membrane Science, 555 (2018) 206–219, https://doi.org/10.1016/j.memsci.2018.03.031

[17] J. Kim, S. Hong, A novel single-pass reverse osmosis configuration for high-purity water production and low energy consumption in seawater desalination, Desalination, 429 (2018) 142–154, https://doi.org/10.1016/j.desal.2017.12.026

[18] W.S. Moore, Large groundwater inputs to coastal waters revealed by 226 Ra enrichments, Nature, 380 (1996) 612–614, https://doi.org/10.1038/380612a0

[19] M.S. Mohsen, O.R. Al-Jayyousi, Brackish water desalination: an alternative for water supply enhancement in Jordan, Desalination, 124 (1999) 163–174, https://doi.org/10.1016/S0011-9164(99)00101-0

[20] I.C. Karagiannis, P.G. Soldatos, Water desalination cost literature: review and assessment, Desalination, 223 (2008) 448–456, https://doi.org/10.1016/j.desal.2007.02.071

[21] M. Elimelech, W.A. Phillip, The future of seawater desalination: energy, technology, and the environment, Science, 333 (2011) 712–717, https://doi.org/10.1126/science.1200488

[22] X. Chen, N.Y. Yip, Unlocking high-salinity desalination with cascading osmotically mediated reverse osmosis: energy and operating pressure analysis, Environmental Science & Technology, 52 (2018) 2242–2250, https://doi.org/10.1021/acs.est.7b05774

[23] L. Henthorne, B. Boysen, State-of-the-art of reverse osmosis desalination pretreatment, Desalination, 356 (2015) 129–139, https://doi.org/10.1016/j.desal.2014.10.039

[24] D.M. Davenport, A. Deshmukh, J.R. Werber, M. Elimelech, High-pressure reverse osmosis for energy-efficient hypersaline brine desalination: current status, design considerations, and research needs, Environmental Science & Technology Letters, 5 (2018) 467–475, https://doi.org/10.1021/acs.estlett.8b00274

[25] C. Boo, R.K. Winton, K.M. Conway, N.Y. Yip, Membrane-less and non-evaporative desalination of hypersaline brines by temperature swing solvent extraction, Environmental Science & Technology Letters, 6 (2019) 359–364, https://doi.org/10.1021/acs.estlett.9b00182

[26] P.V. Brady, R.J. Kottenstette, T.M. Mayer, M.M. Hightower, Inland desalination: challenges and research needs, Journal of Contemporary Water Research & Education, 132 (2005) 46–51, https://doi.org/10.1111/j.1936-704X.2005.mp132001007.x

[27] A. Panagopoulos, K.-J. Haralambous, M. Loizidou, Desalination brine disposal methods and treatment technologies – a review, Science of the Total Environment, 693 (2019) 133545, https://doi.org/10.1016/j.scitotenv.2019.07.351

[28] D.E. López, J.P. Trembly, Desalination of hypersaline brines with joule-heating and chemical pre-treatment: conceptual design and economics, Desalination, 415 (2017) 49–57, https://doi.org/10.1016/j.desal.2017.04.003

[29] J. Kim, J. Kim, J. Kim, S. Hong, Osmotically enhanced dewatering-reverse osmosis (OED-RO) hybrid system: implications for shale gas produced water treatment, Journal of Membrane Science, 554 (2018) 282–290, https://doi.org/10.1016/j.memsci.2018.03.015

[30] J. Kim, J. Kim, J. Lim, S. Lee, C. Lee, S. Hong, Cold-cathode X-ray irradiation pre-treatment for fouling control of reverse osmosis (RO) in shale gas produced water (SGPW) treatment, Chemical Engineering Journal, 374 (2019) 49–58, https://doi.org/10.1016/j.cej.2019.05.158

[31] S. Renou, J. Givaudan, S. Poulain, F. Dirassouyan, P. Moulin, Landfill leachate treatment: review and opportunity, Journal of Hazardous Materials, 150 (2008) 468–493, https://doi.org/10.1016/j.jhazmat.2007.09.077

[32] V. Karanikola, C. Boo, J. Rolf, M. Elimelech, Engineered slippery surface to mitigate gypsum scaling in membrane distillation for treatment of hypersaline industrial wastewaters, Environmental Science & Technology, 52 (2018) 14362–14370, https://doi.org/10.1021/acs.est.8b04836

[33] K. Park, D.Y. Kim, D.R. Yang, Cost-based feasibility study and sensitivity analysis of a new draw solution assisted reverse osmosis (DSARO) process for seawater desalination, Desalination, 422 (2017) 182–193, https://doi.org/10.1016/j.desal.2017.08.026

[34] J. Kim, D.I. Kim, S. Hong, Analysis of an osmotically-enhanced dewatering process for the treatment of highly saline (waste) waters, Journal of Membrane Science, 548 (2018) 685–693, https://doi.org/10.1016/j.memsci.2017.10.048

[35] C.D. Peters, N.P. Hankins, Osmotically assisted reverse osmosis (OARO): five approaches to dewatering saline brines using pressure-driven membrane processes, Desalination, 458 (2019) 1–13, https://doi.org/10.1016/j.desal.2019.01.025

[36] N. Voutchkov, Energy use for membrane seawater desalination – current status and trends, Desalination, 431 (2018) 2–14, https://doi.org/10.1016/j.desal.2017.10.033

[37] Z. Li, R.V. Linares, S. Sarp, G. Amy, Direct and indirect seawater desalination by forward osmosis, in: S. Sarp and N. Hilal (Eds.) Membrane-Based Salinity Gradient Processes for Water Treatment and Power Generation, Elsevier, Amsterdam, 2018, pp. 245–272, https://doi.org/10.1016/C2016-0-02151-4

[38] N. Ghaffour, The challenge of capacity-building strategies and perspectives for desalination for sustainable water use in MENA, Desalination and Water Treatment, 5 (2009) 48–53, https://doi.org/10.5004/dwt.2009.564

[39] J.R. Werber, A. Deshmukh, M. Elimelech, The critical need for increased selectivity, not increased water permeability, for desalination membranes, Environmental Science & Technology Letters, 3 (2016) 112–120, https://doi.org/10.1021/acs.estlett.6b00050

[40] F. Kiand, Supply of desalinated water by the private sector: 30 MGD Singapore seawater desalination plant, in: MEDRC International Conference on Desalination Costing, Conference proceeding, Lemesos, Cyprus, 2004.

[41] S. Lin, M. Elimelech, Staged reverse osmosis operation: configurations, energy efficiency, and application potential, Desalination, 366 (2015) 9–14, https://doi.org/10.1016/j.desal.2015.02.043

[42] G.M. Von Medeazza, "Direct"; and socially-induced environmental impacts of desalination, Desalination, 185 (2005) 57–70, https://doi.org/10.1016/j.desal.2005.03.071

[43] V.G. Gude, Desalination and sustainability – an appraisal and current perspective, Water Research, 89 (2016) 87–106, https://doi.org/10.1016/j.watres.2015.11.012
[44] N. Voutchkov, Seawater desalination – costs and technology trends, in: E.M.V. Hoek and V.V. Tarabara (Eds.) Encyclopedia of Membrane Science and Technology, (2013), John Wiley & Sons, Inc., Hoboken, New Jersey, https://doi.org/10.1002/9781118522318.emst115
[45] A. Shrivastava, S. Rosenberg, M. Peery, Energy efficiency breakdown of reverse osmosis and its implications on future innovation roadmap for desalination, Desalination, 368 (2015) 181–192, https://doi.org/10.1016/j.desal.2015.01.005
[46] M.A. Jamil, B.A. Qureshi, S.M. Zubair, Exergo-economic analysis of a seawater reverse osmosis desalination plant with various retrofit options, Desalination, 401 (2017) 88–98, https://doi.org/10.1016/j.desal.2016.09.032
[47] A. de la Torre, Efficiency optimization in SWRO plant: high efficiency & low maintenance pumps, Desalination, 221 (2008) 151–157, https://doi.org/10.1016/j.desal.2007.02.052
[48] T. Manth, M. Gabor, E. Oklejas Jr, Minimizing RO energy consumption under variable conditions of operation, Desalination, 157 (2003) 9–21, https://doi.org/10.1016/S0011-9164(03)00377-1
[49] A. Subramani, M. Badruzzaman, J. Oppenheimer, J.G. Jacangelo, Energy minimization strategies and renewable energy utilization for desalination: a review, Water Research, 45 (2011) 1907–1920, https://doi.org/10.1016/j.watres.2010.12.032
[50] M. Wilf, C. Bartels, Optimization of seawater RO systems design, Desalination, 173 (2005) 1–12, https://doi.org/10.1016/j.desal.2004.06.206
[51] W. Choi, S. Jeon, S.J. Kwon, H. Park, Y.-I. Park, S.-E. Nam, P.S. Lee, J.S. Lee, J. Choi, S. Hong, Thin film composite reverse osmosis membranes prepared via layered interfacial polymerization, Journal of Membrane Science, 527 (2017) 121–128, https://doi.org/10.1016/j.memsci.2016.12.066
[52] K.P. Lee, T.C. Arnot, D. Mattia, A review of reverse osmosis membrane materials for desalination – development to date and future potential, Journal of Membrane Science, 370 (2011) 1–22, https://doi.org/10.1016/j.memsci.2010.12.036
[53] S.-J. Park, W.-G. Ahn, W. Choi, S.-H. Park, J.S. Lee, H.W. Jung, J.-H. Lee, A facile and scalable fabrication method for thin film composite reverse osmosis membranes: dual-layer slot coating, Journal of Materials Chemistry A, 5 (2017) 6648–6655, https://doi.org/10.1039/C7TA00891K
[54] M. Safarpour, V. Vatanpour, A. Khataee, H. Zarrabi, P. Gholami, M.E. Yekavalangi, High flux and fouling resistant reverse osmosis membrane modified with plasma treated natural zeolite, Desalination, 411 (2017) 89–100, https://doi.org/10.1016/j.desal.2017.02.012
[55] M. Martinez-Mate, B. Martin-Gorriz, V. Martínez-Alvarez, M. Soto-García, J. Maestre-Valero, Hydroponic system and desalinated seawater as an alternative farm-productive proposal in water scarcity areas: energy and greenhouse gas emissions analysis of lettuce production in southeast Spain, Journal of Cleaner Production, 172 (2018) 1298–1310, https://doi.org/10.1016/j.jclepro.2017.10.275
[56] T.P. Hendrickson, M. Bruguera, Impacts of groundwater management on energy resources and greenhouse gas emissions in California, Water Research, 141 (2018) 196–207, https://doi.org/10.1016/j.watres.2018.05.012
[57] D. Voet, J.G. Voet, C.W. Pratt, Fundamentals of Biochemistry: Life at the Molecular Level, John Wiley & Sons, Hoboken, New Jersey, 2016.
[58] R. Tariq, N.A. Sheikh, J. Xamán, A. Bassam, An innovative air saturator for humidification-dehumidification desalination application, Applied Energy, 228 (2018) 789–807, https://doi.org/10.1016/j.apenergy.2018.06.135

[59] T. Pankratz, An overview of seawater intake facilities for seawater desalination. The future of desalination in Texas, Environmental Science, 2 (2004) 1–12, https://texaswater.tamu.edu/readings/desal/seawaterdesal.pdf
[60] D.M. Sigman, M.P. Hain, The biological productivity of the ocean. Nature Education Knowledge, 3 (2012) 21, http://earth-system-biogeochemistry.net/wp-content/uploads/2021/05/Sigman_and_Hain_2012_NatureEdu.pdf
[61] T.M. Missimer, R.G. Maliva, Environmental issues in seawater reverse osmosis desalination: intakes and outfalls. Desalination, 434 (2018) 198–215, https://doi.org/10.1016/j.desal.2017.07.012
[62] N. Voutchkov, SWRO desalination process: on the beach–seawater intakes. Filtration & Separation, 42 (2005) 24–27, https://doi.org/10.1016/S0015-1882(05)70657-1
[63] S.E. De Rijk, J. H. J. M. aivan der, J. G. den Blanken, Bubble size in flotation thickening. Water Research, 28 (1994) 465–473, https://doi.org/10.1016/0043-1354(94)90284-4
[64] P. Scardina, M. Edwards, The fundamentals of bubble formation in water treatment, (2007), http://web.archive.org/web/20060916042031/ http://scholar.lib.vt.edu/theses/available/etd-02242000-20530043/unrestricted/Chapter1.pdf
[65] M.J. Hey, A.M. Hilton, R.D. Bee, The formation and growth of carbon dioxide gas bubbles from supersaturated aqueous solutions. Food Chemistry, 51 (1994) 349–357, https://doi.org/10.1016/0308-8146(94)90185-6
[66] S. Garg, Y. Li, L. Wang, P.M. Schenk, Flotation of marine microalgae: effect of algal hydrophobicity. Bioresource Technology, 121 (2012) 471–474, https://doi.org/10.1016/j.biortech.2012.06.111
[67] J.P. Malley, J.K. Edzwald, Concepts for dissolved-air flotation treatment of drinking waters, Journal of Water Supply: Research and Technology – AQUA, 40 (1991) 7–17.
[68] A. Vlaski, Microcystis Aeruginosa Removal by Dissolved Air Flotation (DAF): Options for Enhanced Process Operation and Kinetic Modelling, (Doctoral dissertation), IHE/TUD, Delft, 1997.
[69] R.K. Henderson, S.A. Parsons, B. Jefferson, Surfactants as bubble surface modifiers in the flotation of algae: dissolved air flotation that utilizes a chemically modified bubble surface, Environmental Science & Technology, 42 (2008) 4883–4888, https://doi.org/10.1021/es702649h
[70] A. B. Alayande, J. Lim, J. Kim, S. Hong, A. S. Al-Amoudi, B. Park, Fouling control in SWRO desalination during harmful algal blooms: a historical review and future developments, Desalination, 543 (2022) 116094, https://doi.org/10.1016/j.desal.2022.116094
[71] R. Henderson, S.A. Parsons, B. Jefferson, The impact of algal properties and pre-oxidation on solid–liquid separation of algae, Water Research, 42 (2008) 1827–1845, https://doi.org/10.1016/j.watres.2007.11.039
[72] T. Pankratz, Red tides close desal plants, Water Desalination Report, 44 (2008).
[73] S. Le Gallou, S. Bertrand, K.H. Madan, Full coagulation and dissolved air flotation: a SWRO key pretreatment step for heavy fouling seawater, in: Proceedings of International Desalination Association World Congress, Perth, Australia, 2011.
[74] D.D. Ratnayaka, Water Supply, Butterworth-Heinemann, Elsevier, Burlington, Massachusetts, 2009.
[75] G. Cha, S. Choi, H. Lee, K. Kim, S. Ahn, S. Hong, Improving energy efficiency of pretreatment for seawater desalination during algal blooms using a novel meshed tube filtration process, Desalination, 486 (2020) 114477, https://doi.org/10.1016/j.desal.2020.114477

[76] M.L. Richlen, S.L. Morton, E.A. Jamali, A. Rajan, D.M. Anderson, The catastrophic 2008–2009 red tide in the Arabian Gulf region, with observations on the identification and phylogeny of the fish-killing dinoflagellate *Cochlodinium polykrikoides*, Harmful Algae, 9 (2010) 163–172, https://doi.org/10.1016/j.hal.2009.08.013

[77] D.M. Anderson, S.F. Boerlage, M.B. Dixon, Harmful Algal Blooms (HABs) and Desalination: a Guide to Impacts, Monitoring and Management, Intergovernmental Oceanographic Commission of UNESCO, 2017, http://dx.doi.org/10.25607/OBP-203

[78] A.R. Guastalli, F.X. Simon, Y. Penru, A. de Kerchove, J. Llorens, S. Baig, Comparison of DMF and UF pre-treatments for particulate material and dissolved organic matter removal in SWRO desalination, Desalination, 322 (2013) 144–150, https://doi.org/10.1016/j.desal.2013.05.005

[79] V. Bonnelye, M.A. Sanz, J.-P. Durand, L. Plasse, F. Gueguen, P. Mazounie, Reverse osmosis on open intake seawater: pre-treatment strategy, Desalination, 167 (2004) 191–200, https://doi.org/10.1016/j.desal.2004.06.128

[80] S.G. Salinas Rodríguez, M.D. Kennedy, J.C. Schippers, G.L. Amy, Organic foulants in estuarine and bay sources for seawater reverse osmosis – Comparing pre-treatment processes with respect to foulant reductions, Desalination and Water Treatment, 9 (2009) 155–164, https://doi.org/10.5004/dwt.2009.766

[81] E. Bar-Zeev, N. Belkin, B. Liberman, I. Berman-Frank, T. Berman, Bioflocculation: chemical free, pre-treatment technology for the desalination industry. Water Research, 47 (2013) 3093–3102, https://doi.org/10.1016/j.watres.2013.03.013

[82] S. Plantier, J.-B. Castaing, N.-E. Sabiri, A. Massé, P. Jaouen, M. Pontié, Performance of a sand filter in removal of algal bloom for SWRO pre-treatment, Desalination and Water Treatment, 51 (2013) 1838–1846, https://doi.org/10.1080/19443994.2012.699336

[83] M. Nair, D. Kumar, Water desalination and challenges: the Middle East perspective: a review, Desalination and Water Treatment, 51 (2013) 2030–2040, https://doi.org/10.1080/19443994.2013.734483

[84] A. Southward, Sea foam, Nature, 172 (1953) 1059–1060, https://doi.org/10.1038/1721059b0

[85] S. Ebrahim, S. Bou-Hamed, M. Abdel-Jawad, N. Burney, Microfiltration system as a pretreatment for RO units: technical and economic assessment, Desalination, 109 (1997) 165–175, https://doi.org/10.1016/S0011-9164(97)00062-3

[86] W. Ma, Y. Zhao, L. Wang, The pretreatment with enhanced coagulation and a UF membrane for seawater desalination with reverse osmosis, Desalination, 203 (2007) 256–259, https://doi.org/10.1016/j.desal.2006.02.020

[87] E. Fontananova, M. Bahattab, S. Aljlil, M. Alowairdy, G. Rinaldi, D. Vuono, J. Nagy, E. Drioli, G. Di Profio, From hydrophobic to hydrophilic polyvinylidenefluoride (PVDF) membranes by gaining new insight into material's properties, RSC Advances, 5 (2015) 56219–56231, https://doi.org/10.1039/C5RA08388E

[88] G. Pearce, S. Talo, K. Chida, A. Basha, A. Gulamhusein, Pretreatment options for large scale SWRO plants: case studies of OF trials at Kindasa, Saudi Arabia, and conventional pretreatment in Spain, Desalination, 167 (2004) 175–189, https://doi.org/10.1016/j.desal.2004.06.127

[89] D.F. Halpern, J. McArdle, B. Antrim, UF pretreatment for SWRO: pilot studies, Desalination, 182 (2005) 323–332, https://doi.org/10.1016/j.desal.2005.02.031

[90] R.W. Baker, Membrane Technology and Applications, John Wiley & Sons, Chichester, United Kingdom, 2012.

[91] A. Hassan, M. Al-Sofi, A. Al-Amoudi, A. Jamaluddin, A. Farooque, A. Rowaili, A. Dalvi, N. Kither, G. Mustafa, I. Al-Tisan, A new approach to membrane and

thermal seawater desalination processes using nanofiltration membranes (Part 1), Desalination, 118 (1998) 35–51, https://doi.org/10.1016/S0011-9164(98)00079-4

[92] Y. Choi, J. Kweon, D. Kim, S. Lee, Evaluation of various pretreatment for particle and inorganic fouling control on performance of SWRO, Desalination, 247 (2009) 137–147, https://doi.org/10.1016/j.desal.2008.12.019

[93] A.H. Alshahri, L. Fortunato, N. Ghaffour, T. Leiknes, Advanced coagulation using in-situ generated liquid ferrate, Fe (VI), for enhanced pretreatment in seawater RO desalination during algal blooms, Science of the Total Environment, 685 (2019) 1193–1200, https://doi.org/10.1016/j.scitotenv.2019.06.286

[94] J. Liu, K. He, S. Tang, T. Wang, Z. Zhang, A comparative study of ferrous, ferric and ferrate pretreatment for ceramic membrane fouling alleviation in reclaimed water treatment, Separation and Purification Technology, 217 (2019) 118–127, https://doi.org/10.1016/j.seppur.2019.01.040

[95] S.A. Urrea, F.D. Reyes, B.P. Suárez, A. Juan, Technical review, evaluation and efficiency of energy recovery devices installed in the Canary Islands desalination plants, Desalination, 450 (2019) 54–63, https://doi.org/10.1016/j.desal.2018.07.013

[96] J. Schallenberg-Rodríguez, J.M. Veza, A. Blanco-Marigorta, Energy efficiency and desalination in the Canary Islands, Renewable and Sustainable Energy Reviews, 40 (2014) 741–748, https://doi.org/10.1016/j.rser.2014.07.213

[97] K. Park, P.A. Davies, A compact hybrid batch/semi-batch reverse osmosis (HBSRO) system for high-recovery, low-energy desalination, Desalination, 504 (2021) 114976, https://doi.org/10.1016/j.desal.2021.114976

[98] K. Park, L. Burlace, N. Dhakal, A. Mudgal, N.A. Stewart, P.A. Davies, Design, modelling and optimisation of a batch reverse osmosis (RO) desalination system using a free piston for brackish water treatment, Desalination, 494 (2020) 114625, https://doi.org/10.1016/j.desal.2020.114625

[99] Z. Gal, A. Efraty, CCD series no. 18: record low energy in closed-circuit desalination of ocean seawater with nanoH_2O elements without ERD, Desalination and Water Treatment, 57 (2016) 9180–9189, https://doi.org/10.1080/19443994.2015.1035500

[100] A. Nemdili, D.-H. Hellmann, The requirements to successful centrifugal pump application for desalination and power plant processes, Desalination, 126 (1999) 199–205, https://doi.org/10.1016/S0011-9164(99)00175-7

[101] N. Sawaki, C.-L. Chen, Cost evaluation for a two-staged reverse osmosis and pressure retarded osmosis desalination process, Desalination, 497 (2021) 114767, https://doi.org/10.1016/j.desal.2020.114767

[102] N. Voutchkov, Desalination Engineering: Planning and Design, McGraw-Hill Professional, New York, New York, 2012.

[103] J. Kim, K. Park, S. Hong, Application of two-stage reverse osmosis system for desalination of high-salinity and high-temperature seawater with improved stability and performance, Desalination, 492 (2020) 114645, https://doi.org/10.1016/j.desal.2020.114645

[104] J. Kim, K. Park, S. Hong, Optimization of two-stage seawater reverse osmosis membrane processes with practical design aspects for improving energy efficiency, Journal of Membrane Science, 601 (2020) 117889, https://doi.org/10.1016/j.memsci.2020.117889

[105] E. Spiritos, C. Lipchin, Desalination in Israel, in, Springer, Dordrecht, Netherlands, 2013, pp. 101–123.

[106] B. Sauvet-Goichon, Ashkelon desalination plant – a successful challenge, Desalination, 203 (2007) 75–81, https://doi.org/10.1016/j.desal.2006.03.525

[107] M. Faigon, Y. Egozy, D. Hefer, M. Ilevicky, Y. Pinhas, Hadera desalination plant two years of operation, Desalination and Water Treatment, 51 (2013) 132–139, https://doi.org/10.1080/19443994.2012.701414

[108] J. Kim, S. Hong, Optimizing seawater reverse osmosis with internally staged design to improve product water quality and energy efficiency, Journal of Membrane Science, 568 (2018) 76–86, https://doi.org/10.1016/j.memsci.2018.09.046

[109] B. Peñate, L. García-Rodríguez, Reverse osmosis hybrid membrane inter-stage design: a comparative performance assessment, Desalination, 281 (2011) 354–363, https://doi.org/10.1016/j.desal.2011.08.010

[110] N. Nada, T. Attenborough, Y. Ito, Y. Maeda, K. Tokunaga, H. Iwahashi, SWRO drinking water project in Shuqaiq: advanced BWRO, membrane oxidation, and scaling, IDA Journal of Desalination and Water Reuse, 3 (2011) 30–39, https://doi.org/10.1179/ida.2011.3.4.30

[111] W.H. Organization, Safe Drinking-Water from Desalination, in World Health Organization, Geneva, Switzerland, 2011.

[112] J.S. Taylor, J.D. Dietz, A.A. Randall, S.K. Hong, C.D. Norris, L.A. Mulford, J.M. Arevalo, S. Imran, M. Le Puil, S. Liu, I. Mutoti, J. Tang, W. Xiao, C. Cullen, R. Heaviside, A. Mehta, M. Patel, F. Vasquez, D. Webb, Effects of Blending on Distribution Water Quality, American Water Works Association Research Foundation, Denver, CO, 2005.

[113] A. Withers, Options for recarbonation, remineralisation and disinfection for desalination plants, Desalination, 179 (2005) 11–24, https://doi.org/10.1016/j.desal.2004.11.051

[114] G. Mauguin, P. Corsin, Concentrate and other waste disposals from SWRO plants: characterization and reduction of their environmental impact, Desalination, 182 (2005) 355–364, https://doi.org/10.1016/j.desal.2005.02.033

[115] R. Maliva, T. Missimer, Environmental issues in desalination, in: R. Maliva and T. Missimer (Eds.) Arid Lands Water Evaluation and Management, Springer, Berlin (Heidelberg), 2012, pp. 749–769, https://doi.org/10.1007/978-3-642-29104-3

[116] P.M.G. Olabarria, Constructive Engineering of Large Reverse Osmosis Desalination Plants, Chemical Publishing Company, USA, 2015.

[117] M. Abualtayef, H. Al-Najjar, Y. Mogheir, A.K. Seif, Numerical modeling of brine disposal from Gaza central seawater desalination plant. Arabian Journal of Geosciences, 9 (2016) 1–18, https://doi.org/10.1007/s12517-016-2591-7

[118] E. Mackey, T. Seacord, Final Report, Regional Solution for Concentration Management, (2008).

[119] C. Duclos-Orsello, W. Li, C.-C. Ho, A three mechanism model to describe fouling of microfiltration membranes, Journal of Membrane Science, 280 (2006) 856–866, https://doi.org/10.1016/j.memsci.2006.03.005

[120] W. Yuan, A. Kocic, A.L. Zydney, Analysis of humic acid fouling during microfiltration using a pore blockage–cake filtration model, Journal of Membrane Science, 198 (2002) 51–62, https://doi.org/10.1016/S0376-7388(01)00622-6

[121] J.G. Wijmans, R.W. Baker, The solution-diffusion model: a review, Journal of Membrane Science, 107 (1995) 1–21, https://doi.org/10.1016/0376-7388(95)00102-I

[122] H.-J. Oh, T.-M. Hwang, S. Lee, A simplified simulation model of RO systems for seawater desalination, Desalination, 238 (2009) 128–139, https://doi.org/10.1016/j.desal.2008.01.043

[123] M. Cheryan, Ultrafiltration and Microfiltration Handbook, CRC Press, Boca Raton, Florida, 1998.

[124] G.M. Geise, H.B. Park, A.C. Sagle, B.D. Freeman, J.E. McGrath, Water permeability and water/salt selectivity tradeoff in polymers for desalination, Journal of Membrane Science, 369 (2011) 130–138, https://doi.org/10.1016/j.memsci.2010.11.054

[125] N.Y. Yip, M. Elimelech, Performance limiting effects in power generation from salinity gradients by pressure retarded osmosis, Environmental Science & Technology, 45 (2011) 10273–10282, https://doi.org/10.1021/es203197e

[126] S. Kim, E.M. Hoek, Modeling concentration polarization in reverse osmosis processes, Desalination, 186 (2005) 111–128, https://doi.org/10.1016/j.desal.2005.05.017

[127] I. Sutzkover, D. Hasson, R. Semiat, Simple technique for measuring the concentration polarization level in a reverse osmosis system, Desalination, 131 (2000) 117–127, https://doi.org/10.1016/S0011-9164(00)90012-2

[128] M. Qasim, M. Badrelzaman, N.N. Darwish, N.A. Darwish, N. Hilal, Reverse osmosis desalination: a state-of-the-art review, Desalination, 459 (2019) 59–104, https://doi.org/10.1016/j.desal.2019.02.008

[129] P. Luis, Fundamental Modeling of Membrane Systems: Membrane and Process Performance, Elsevier, Amsterdam, Netherlands, 2018.

[130] J.R. Welty, C.E. Wicks, G. Rorrer, R.E. Wilson, Fundamentals of Momentum, Heat, and Mass Transfer, John Wiley & Sons, Hoboken, New Jersey, 2009.

[131] S. Sablani, M. Goosen, R. Al-Belushi, M. Wilf, Concentration polarization in ultrafiltration and reverse osmosis: a critical review, Desalination, 141 (2001) 269–289, https://doi.org/10.1016/S0011-9164(01)85005-0

[132] C.P. Koutsou, S.G. Yiantsios, A.J. Karabelas, A numerical and experimental study of mass transfer in spacer-filled channels: effects of spacer geometrical characteristics and Schmidt number, Journal of Membrane Science, 326 (2009) 234–251, https://doi.org/10.1016/j.memsci.2008.10.007

[133] M. Shibuya, M. Yasukawa, S. Goda, H. Sakurai, T. Takahashi, M. Higa, H. Matsuyama, Experimental and theoretical study of a forward osmosis hollow fiber membrane module with a cross-wound configuration, Journal of Membrane Science, 504 (2016) 10–19, https://doi.org/10.1016/j.memsci.2015.12.040

[134] E.M. Hoek, A.S. Kim, M. Elimelech, Influence of crossflow membrane filter geometry and shear rate on colloidal fouling in reverse osmosis and nanofiltration separations, Environmental Engineering Science, 19 (2002) 357–372, https://doi.org/10.1089/109287502320963364

[135] C.O. Bennett, J.E. Myers, Momentum, Heat, and Mass Transfer, McGraw-Hill, New York, 1982.

[136] T. Ishigami, H. Matsuyama, Numerical modeling of concentration polarization in spacer-filled channel with permeation across reverse osmosis membrane, Industrial & Engineering Chemistry Research, 54 (2015) 1665–1674, https://doi.org/10.1021/ie5039665

[137] R. Deissler, Recent Advances in Heat and Mass Transfer, JP Hartnett, ed., McGraw-Hill Press, New York, 1961, p. 253.

[138] R.H. Notter, C. Sleicher, The eddy diffusivity in the turbulent boundary layer near a wall, Chemical Engineering Science, 26 (1971) 161–171, https://doi.org/10.1016/0009-2509(71)86088-8

[139] W. Pinczewski, S. Sideman, A model for mass (heat) transfer in turbulent tube flow. Moderate and high Schmidt (Prandtl) numbers, Chemical Engineering Science, 29 (1974) 1969–1976, https://doi.org/10.1016/0009-2509(74)85016-5

[140] A. Achilli, T.Y. Cath, A.E. Childress, Power generation with pressure retarded osmosis: an experimental and theoretical investigation, Journal of Membrane Science, 343 (2009) 42–52, https://doi.org/10.1016/j.memsci.2009.07.006

[141] M. Sekino, Mass transfer characteristics of hollow fiber RO modules, Journal of Chemical Engineering of Japan, 28 (1995) 843–846, https://doi.org/10.1252/jcej.28.843

[142] A. Kumano, H. Matsuyama, Analysis of hollow fiber reverse osmosis membrane module of axial flow type, Journal of Applied Polymer Science, 123 (2012) 463–471, https://doi.org/10.1002/app.34404
[143] N. Voutchkov, Pretreatment for Reverse Osmosis Desalination, Elsevier, Amsterdam, Netherlands, 2017.
[144] G.P.S. Ibrahim, A.M. Isloor, R. Farnood, Reverse Osmosis Treatment Techniques, Fouling, and Control Strategies, Elsevier, Amsterdam, Netherlands, 2020.
[145] C.Y. Tang, T.H. Chong, A.G. Fane, Colloidal interactions and fouling of NF and RO membranes: a review. Advances in Colloid and Interface Science, 164 (2011) 126–143, https://doi.org/10.1016/j.cis.2010.10.007
[146] R. Singh, N. Hankins, Emerging Membrane Technology for Sustainable Water Treatment, Elsevier, Amsterdam, Netherlands, 2016.
[147] T.A. Saleh, V.K. Gupta, Nanomaterial and Polymer Membranes: Synthesis, Characterization, and Applications, Elsevier, Amsterdam, Netherlands, 2016.
[148] M. Qasim, M. Badrelzaman, N.N. Darwish, N.A. Darwish, N. Hilal, Reverse osmosis desalination: a state-of-the-art review, Desalination, 459 (2019) 59–104, https://doi.org/10.1016/j.desal.2019.02.008
[149] A.F. Isamil, K.C. Khulbe, T. Matsuura, RO membrane fouling. In: Reverse Osmosis. Elsevier, 2019, pp. 189–220, https://doi.org/10.1016/B978-0-12-811468-1.00008-6
[150] A. Matin, Z. Khan, S.M. J. Zaidi, M.C. Boyce, Biofouling in reverse osmosis membranes for seawater desalination: phenomena and prevention. Desalination, 281 (2011) 1–16, https://doi.org/10.1016/j.desal.2011.06.063
[151] O. Sánchez, Microbial diversity in biofilms from reverse osmosis membranes: a short review. Journal of Membrane Science, 545 (2018) 240–249, https://doi.org/10.1016/j.memsci.2017.09.082
[152] C. Xu, W-C. Chin, P. Lin, H. Chen, M-H. Chiu, D. C. Waggoner, W. Xing, L. Sun, K. A. Schwehr, P.G. Hatcher, A. Quigg, P.H. Santschi, Comparison of microgels, extracellular polymeric substances (EPS) and transparent exopolymeric particles (TEP) determined in seawater with and without oil, Marine Chemistry, 215 (2019) 103667, https://doi.org/10.1016/j.marchem.2019.103667
[153] J.B. Kaplan, Biofilm dispersal: mechanisms, clinical implications, and potential therapeutic uses. Journal of Dental Research, 89 (2010) 205–218, https://doi.org/10.1177/0022034509359403
[154] J. MacAdam, S.A.J.R.V.I.E.S. Parsons, Calcium carbonate scale formation and control, Reviews in Environmental Science and Bio/Technology, 3 (2004) 159–169, https://doi.org/10.1007/s11157-004-3849-1
[155] M.H. Al-Khaldi, A. AlJuhani, S.H. Al-Mutairi, M.N. Gurmen, New insights into the removal of calcium sulfate scale, in: SPE European Formation Damage Conference, Society of Petroleum Engineers, 2011.
[156] S. Bhattacharjee, Concentration Polarization: Early Theories, in, Water Planet, Los Angeles, California, 2017.
[157] D.Y. Kim, B. Gu, D.R. Yang, An explicit solution of the mathematical model for osmotic desalination process, Korean Journal of Chemical Engineering, 30 (2013) 1691–1699, https://doi.org/10.1007/s11814-013-0123-7
[158] K. Park, D.Y. Kim, Y.H. Jang, M.-G. Kim, D.R. Yang, S. Hong, Comprehensive analysis of a hybrid FO/crystallization/RO process for improving its economic feasibility to seawater desalination, Water Research, 171 (2020) 115426, https://doi.org/10.1016/j.watres.2019.115426
[159] D.Y. Kim, B. Gu, J.H. Kim, D.R. Yang, Theoretical analysis of a seawater desalination process integrating forward osmosis, crystallization, and reverse osmosis, Journal of Membrane Science, 444 (2013) 440–448, https://doi.org/10.1016/j.memsci.2013.05.035

[160] D. Cohen-Tanugi, R.K. McGovern, S.H. Dave, J.H. Lienhard, J.C. Grossman, Quantifying the potential of ultra-permeable membranes for water desalination, Energy Environmental Science, 7 (2014) 1134–1141, https://doi.org/10.1039/C3EE43221A

[161] R.L. Stover, Seawater reverse osmosis with isobaric energy recovery devices, Desalination, 203 (2007) 168–175, https://doi.org/10.1016/j.desal.2006.03.528

[162] S.R. Osipi, A.R. Secchi, C.P. Borges, Cost analysis of forward osmosis and reverse osmosis in a case study, Current Trends and Future Developments on (Bio-) Membranes: Reverse and Forward Osmosis: Principles, Applications, Advances, Elsevier, Amsterdam, Netherlands, 2019, p. 305.

[163] T. Heskes, Practical confidence and prediction intervals, in: Advances in Neural Information Processing Systems, The MIT Press, Cambridge, Massachusetts, 1997, pp. 176–182.

[164] R. Dybowski, V. Gant, Clinical Applications of Artificial Neural Networks, Cambridge University Press, Cambridge, United Kingdom, 2001.

[165] J. Fox, G. Monette, An R and S-Plus Companion to Applied Regression, Sage Publications, Thousand Oaks, California, 2002.

[166] R. Stover, J. Martin, Titan PX-1200 Energy recovery device – test results from the Inima Los Cabos, Mexico, seawater RO facility, Desalination and Water Treatment, 3 (2009) 179–182, https://doi.org/10.5004/dwt.2009.458

[167] L.A. Hoover, W.A. Phillip, A. Tiraferri, N.Y. Yip, M. Elimelech, Forward with osmosis: emerging applications for greater sustainability, Environmental Science & Technology, 45 (2011) 9824–9830, https://doi.org/10.1021/es202576h

[168] Los Cabos, (IDAM de Los Cabos, http://www.inima.es/es/idam_cabos_mexico, (accessed September 2018)).

[169] G. Wetterau, Desalination of Seawater: M61, American Water Works Association, Denver, Colorado, 2011.

[170] M.L. Stone, C. Rae, F.F. Stewart, A.D. Wilson, Switchable polarity solvents as draw solutes for forward osmosis, Desalination, 312 (2013) 124–129, https://doi.org/10.1016/j.desal.2012.07.034

[171] IDA Desalination Yearbook 2016–2017, Media Analytics Ltd., Oxford, 2017.

[172] R.B. A., G. Olivares, Buenas prácticas y uso eficiente de agua en la industria minera, Comisión Chilena del Cobre (Cochilo), Santiago, 2008.

[173] M. Petry, M.A. Sanz, C. Langlais, V. Bonnelye, J.-P. Durand, D. Guevara, W.M. Nardes, C.H. Saemi, The El Coloso (Chile) reverse osmosis plant, Desalination, 203 (2007) 141–152, https://doi.org/10.1016/j.desal.2006.05.007

[174] S. Vicuña, in, Nexo Agua y Energía en Chile, https://cambioglobal.uc.cl/images/presentaciones/Seminarios_Extension/Nexo_Agua_Energia/SVicuna-Nexo-Agua-Energia-Chile.pdf (accessed September 2018).

[175] D.M. Anderson, S.F.E. Boerlage, M.B. Dixon, Harmful Algal Blooms (HABs) and Desalination: A Guide to Impacts, Monitoring and Management, UNESCO, Paris, France, 2017.

[176] R.L. Burk, M.B. Dixon, D. Kim-Hak, P.M. Vega, A comparison of three reverse osmosis membranes at La Chimba desalination plant, Antofagasta, Chile, in: International Desalination Association World Congress on Desalination and Water Reuse 2013, Tianjin, October, 2013.

[177] M. Trifkovic, M. Sheikhzadeh, S. Rohani, Multivariable real-time optimal control of a cooling and antisolvent semibatch crystallization process, AIChE Journal, 55 (2009) 2591–2602, https://doi.org/10.1002/aic.11868

[178] F.A. Marchena, Efficiency improvement of seawater desalination processes: The case of the WEB Aruba NV on the island of Aruba (PhD thesis, University of Twente, 2013).

[179] P. Juif, Reverse osmosis: case study Aruba (Reverse osmosis: case study Aruba, https://report.nat.gov.tw/ReportFront/PageSystem/reportFileDownload/C10002957/005 (accessed September 2018)).
[180] S. Meyer-Steele, A.V. Gottberg, J.L. Talavera, Seawater Reverse Osmosis Plants in the Caribbean Recover Energy and Brine and Reduce Costs, SUEZ Water Technologies & Solutions, N/A, 2017.
[181] Michael Brand, Reducing the dependence of water supply systems on reliable rainfall patterns, https://www.churchilltrust.com.au/media/fellows/Brand_Michael_2006.pdf (accessed September 2018).
[182] N. Voutchkov, Chapter 1 – membrane desalination – process selection, design, and implementation, in: V.G. Gude (Ed.) Sustainable Desalination Handbook, Butterworth-Heinemann, Oxford, United Kingdom, 2018, pp. 3–24.
[183] J. Cabero García, Proceso de desalación de agua de mar mediante un sistema de osmosis inversa de muy alta conversión en tres etapas con recirculación de permeado y doble sistema de recuperación de energía (PhD thesis, University of the Basque Country, 2016).
[184] J.R. McCutcheon, R.L. McGinnis, M. Elimelech, A novel ammonia – carbon dioxide forward (direct) osmosis desalination process, Desalination, 174 (2005) 1–11, https://doi.org/10.1016/j.desal.2004.11.002
[185] X. Zheng, D. Chen, Q. Wang, Z. Zhang, Seawater desalination in China: retrospect and prospect, Chemical Engineering Journal, 242 (2014) 404–413, https://doi.org/10.1016/j.cej.2013.12.104
[186] J.-S. Liu, S.-L. Pang, Design outline of sea water desalination project for Huaneng Yuhuan power plant, Technology of Water Treatment, 31 (2013) 73–75.
[187] J. R. Wang, Introduction to seawater desalination project of Huaneng Yuhuan power plant, Electric Power Construction, 2 (2008) 015.
[188] T. Hamano, Operation of Fukuoka sea water desalination plant, Bulletin of the Society of Sea Water Science, Japan, 60 (2006) 415–421.
[189] F. Nobuya, M. Hideto, High recovery system in seawater reverse osmosis plants, Journal of Applied Polymer Science, 108 (2008) 3403–3410, https://doi.org/10.1002/app.27943
[190] Y. Kim, S. Ihm, S. Woo, Improving energy efficiency of a seawater reverse osmosis plant in Gijang, Busan, Korea, in: 10th International Desalination Workshop (IDW 2017), Busan, November, 2017.
[191] M.-H. Hwang, I.S. Kim, Comparative analysis of seawater desalination technology in Korea and overseas, Journal of Korean Society of Environmental Engineers, 38 (2016) 255–268, https://doi.org/10.4491/KSEE.2016.38.5.255
[192] S. Woo, Y. Kim, Reduction of power consumption by variable speed operation of high pressure pump in seawater reverse osmosis desalination plant, The KSFM Journal of Fluid Machinery, 20 (2017) 33–39, https://doi.org/10.5293/kfma.2017.20.5.033
[193] J. Blesing, C. Pelekani, Seawater desalination: a sustainable solution to world water shortage, Water e-Journal, 1 (2016) 1–8, https://doi.org/10.21139/wej.2016.003
[194] Australian Water Quality Centre, Adelaide Desalination Project, Australian Water Quality Centre, Adelaide, 2013.
[195] G.J. Crisp, Desalination in Australia. Sustainably drought proofing Australia, in, Seminar in Chemical Panel Engineers Australia WA Division & The Institution of Chemical Engineers (WA), Water Conditioning & Purification International Magazine, Perth, March, 2011, https://wcponline.com/2011/11/17/desalination-sustainably-drought-proofing-australia/
[196] Gold Coast Desalination Plant, Queensland, https://www.water-technology.net/projects/gold-coast-plant/ (accessed September 2018).

[197] M.A. Sanz, D. Zarzo, Australian desalination plan: PPP models & solutions for the environment 2 case studies, https://www.cmimarseille.org/sites/default/files/newsite/library/files/en/2.10.%20D.%20Sarzo%3B%20M.%20Sanz_%20Australian%20Desalination%20Plants_%20PPP%20Models%20and%20Solutions%20for%20the%20Environment-ilovepdf-compressed.pdf (accessed September 2018).
[198] Victorian Desalination Plant, https://www.aquasure.com.au/uploads/files/ENV-000-PL-001%20-%20Environmental%20Management%20Plan%20(web%20version)-1434948886.pdf (accessed September 2018).
[199] Reverse osmosis: membranes help beat the drought, Filtration & Separation, 46 (2009) 23–24, https://doi.org/10.1016/S0015-1882(09)70055-2
[200] J.R. McCutcheon, M. Elimelech, Influence of concentrative and dilutive internal concentration polarization on flux behavior in forward osmosis, Journal of Membrane Science, 284 (2006) 237–247, https://doi.org/10.1016/j.memsci.2006.07.049
[201] D. Bosman, Large scale desalination projects in Australia: an overview and lessons learned, in: International Conference on Fresh Water Governance for Sustainable Development, Drakensberg, November, 2012.
[202] W. Fleischmann, A. Mersmann, Solubility, density and viscosity for sodium sulfate-methanol-water systems, Journal of Chemical & Engineering Data, 29 (1984) 452–456, https://doi.org/10.1021/je00038a026
[203] T.R. EnergÏa, in, International EPC Contractor Presentation, N/A, April, 2017.
[204] R. Singh, Membrane Technology and Engineering for Water Purification: Application, Systems Design and Operation, Butterworth-Heinemann, Oxford, United Kingdom, 2014.
[205] Sydney Desalination Plant, https://www.acaa.net.au/wp-content/uploads/2015/05/ACAA-submission-Sydney-Desalination-Plant.pdf (accessed September 2018).
[206] K.K. Mehrotra, Desalination: the quest to quench India's thirst for drinking water, in: Technical Talk on "De-Salination Technology", New Delhi, December, 2015.
[207] Chennai Metropolitan Water Supply and Sewerage Board (CMWSSB), Construction and Operation & Maintenance for 7 Years of 100 MLD Seawater Reverse Osmosis Desalination Plant at Nemmeli, Tamilnadu, 2010.
[208] VA Tech Wabag FY 13 results update, http://www.wabag.com/wp-content/uploads/2018/07/9681_740_Analyst2bPresentation2b-2bQ42b-F13.pdf (accessed September 2018).
[209] A. Efraty, Closed circuit desalination series no-6: conventional RO compared with the conceptually different new closed circuit desalination technology, Desalination and Water Treatment, 41 (2012) 279–295, https://doi.org/10.1080/19443994.2012.664741
[210] C. Huttche, R. Orquiza, Y.P. Lelean, Pollution Control Study for Tuas Desalination and Power Plant Project, Environmental Professionals, Singapore, 2011.
[211] A. Cipollina, G. Micale, L. Rizzuti, Seawater Desalination: Conventional and Renewable Energy Processes, Springer Science & Business Media, Heidelberg, Germany, 2009.
[212] Hyflux, Tuaspring Integrated Water and Power Project, Hyflux, Singapore, n.d.
[213] G. Blandin, A.R. Verliefde, C.Y. Tang, P. Le-Clech, Opportunities to reach economic sustainability in forward osmosis–reverse osmosis hybrids for seawater desalination, Desalination, 363 (2015) 26–36, https://doi.org/10.1016/j.desal.2014.12.011
[214] Sadara SWRO Plant- Saudi Arabia, http://www.sidem-desalination.com/main-references/case-studies/Sadara_SWRO.htm (accessed September 2017).
[215] C. Tan, H. Ng, A novel hybrid forward osmosis-nanofiltration (FO-NF) process for seawater desalination: draw solution selection and system configuration, Desalination and Water Treatment, 13 (2010) 356–361, https://doi.org/10.5004/dwt.2010.1733

[216] M.A. Sanza, V. Bonnélyea, G. Cremerb, Fujairah reverse osmosis plant: 2 years of operation, Desalination, 203 (2007) 91–99, https://doi.org/10.1016/j.desal.2006.03.526

[217] D. Cheng, W. Gong, N. Li, Response surface modeling and optimization of direct contact membrane distillation for water desalination, Desalination, 394 (2016) 108–122, https://doi.org/10.1016/j.desal.2016.04.029

[218] I. Hitsov, T. Maere, K. De Sitter, C. Dotremont, I. Nopens, Modelling approaches in membrane distillation: a critical review, Separation and Purification Technology, 142 (2015) 48–64, https://doi.org/10.1016/j.seppur.2014.12.026

[219] Layyah & Khorfakhan SWRO Plants Run with DOW FILMTEC™ Technology, Dow Water & Process Solutions, Dow Chemical, US, N/A.

[220] C. Gasson, C. Gonzalez-Manchon, F. Alvarado-Revilla, Global Water Intelligence, Desalination markets 2010, Media Analytics Ltd, Oxford, 2010.

[221] IDA Desalination Yearbook 2008–2009, Media Analytics Ltd, Oxford, 2009.

[222] D.E. Lindeman, N. Callahan, Tampa bay seawater desalination DBOOT project: lessons learned, in: 10th Annual CWWA Conference and Exhibition, Grand Cayman, October, 2001.

[223] N. Voutchkov, in, Presented in Part at the Training Course on Modeling the Cost of Desalination, Muscat, June, 2013.

[224] G. Gude, Sustainable Desalination Handbook: Plant Selection, Design and Implementation, Elsevier Science, Oxford, United Kingdom, 2018.

[225] N. Melián-Martel, J.J. Sadhwani, S. Malamis, M. Ochsenkühn-Petropoulou, Structural and chemical characterization of long-term reverse osmosis membrane fouling in a full scale desalination plant, Desalination, 305 (2012) 44–53, https://doi.org/10.1016/j.desal.2012.08.011

[226] B. Peñate, L. García-Rodríguez, Energy optimisation of existing SWRO (seawater reverse osmosis) plants with ERT (energy recovery turbines): technical and thermoeconomic assessment, Energy, 36 (2011) 613–626, https://doi.org/10.1016/j.energy.2010.09.056

[227] R. Lemes, R. Falcon, R. Arocha, J. Curbelo, V. Platas, L.D. Lorenzo, Different designs in energy savings of SWRO plant of Las Palmas III, Desalination and Water Treatment, 51 (2013) 749–758, https://doi.org/10.1080/19443994.2012.699342

[228] O.L. Villa Sallangos, Operating experience of the Dhekelia seawater desalination plant using an innovative energy recovery system, Desalination, 173 (2005) 91–102, https://doi.org/10.1016/j.desal.2004.07.045

[229] M. Schenkeveld, R. Morris, B. Budding, J. Helmer, S. Innanen, Seawater and Brackish Water Desalination in the Middle East, North Africa and Central Asia – A Review of Key Issues and Experience in Six Countries (Algeria, Tunisia, Jordan, Uzbekistan, Malta, Cyprus), World Bank, US, 2004.

[230] E. Koutsakos, D. Moxey, More water, less energy and reduced CO_2 emissions – the Larnaca desalination plant, Desalination and Water Treatment, 5 (2009) 68–73, https://doi.org/10.5004/dwt.2009.561

[231] C. Bartels, S. Cioffi, S. Rybar, M. Wilf, E. Koutsakos, Long term experience with membrane performance at the Larnaca desalination plant, Desalination, 221 (2008) 92–100, https://doi.org/10.1016/j.desal.2007.01.070

[232] F. Lopez, M. Araus, R. Gimenez, E. Munoz, G. Romero, J.M. Bastida, Energy consumption optimization in a large seawater desalination plant (Aguilas, 210,000 m^3/day), in: International Desalination Association World Congress on Desalination and Water Reuse 2015, San Diego, 2015.

[233] R.B. Candel, C.G.G. Soto, D.Z. Martínez, A.D. Pérez, M.G.M. Pfeiffer, Á. García, Design of the Pretreatment at the New Águilas-Guadalentín Desalination Plant in: IDA World Congress 2009, Atlantis, November, 2009.
[234] acuaMed, Aguilas/Guadalentin Desalination Plant, http://www.acuamed.es/media/actuaciones/88/aguilas-corta-def.pdf (accessed September 2018).
[235] O. Alvarado, Planta Desaladora Çguilas/GuadalentÓn, in: Ii Seminario Internacional DesalaciÕn En Antofagasta, Antofagasta, 2012.
[236] J.O. Cantos, A.M.R. Amorós, Libro jubilar en Homenaje al Profesor Antonio Gil Olcina. Versión ampliada, Publicaciones de la Universidad de Alicante, San Vicente del Raspeig, Spain, 2016.
[237] E. Drioli, A. Criscuoli, E. Curcio, Integrated membrane operations for seawater desalination, Desalination, 147 (2002) 77–81, https://doi.org/10.1016/S0011-9164(02)00579-9
[238] S. Vigneswaran, Waste Water Treatment Technologies – Volume III, EOLSS Publ., Oxford, United Kingdom, 2009.
[239] M.A. Sanz, C. Miguel, The role of SWRO Barcelona-Llobregat Plant in the water supply system of Barcelona Area, Desalination and Water Treatment, 51 (2013) 111–123, https://doi.org/10.1080/19443994.2012.699250
[240] B. Marilhet, Barcelona Desalination Plant, A Case Study on Water Technologies in Spain, Suez Environnement, France, 2009.
[241] in, IDA Desalination Yearbook 2009–2010. Media Analytics Ltd, Oxford, 2010.
[242] acuaMed, Desalination plant in Campo de Dalias (Almeria), http://www.acuamed.es/media/actuaciones/171/campodalias-corta-def.pdf (accessed September 2018).
[243] Campo de Dalías Desalination Plant (Almeria), FuturENVIRO, Madrid, 2015.
[244] Descripción del proyecto y mejoras de la desaladora de agua de mar del Campo de Dalías, https://www.iagua.es/blogs/veolia-water/descripcion-del-proyecto-y-mejoras-de-la-desaladora-de-agua-de-mar-del-campo-de-dalias (accessed September 2018).
[245] Carboneras SWRO, http://www.acciona-agua.com/areas-of-activity/projects/dc-water-treatment-plants/swro/carboneras/ (accessed September 2018).
[246] acuaMed, Carboneras Desalination Plant (Almeria), http://www.acuamed.es/media/actuaciones/101/carboneras-corta-def.pdf (accessed September 2018).
[247] M.A. García-Rubio, J. Guardiola, Desalination in Spain: a growing alternative for water supply, International Journal of Water Resources Development, 28 (2012) 171–186, https://doi.org/10.1080/07900627.2012.642245
[248] Global Water Intelligence (GWI), DesalData, Global Water Intelligence (GWI), Oxford, United Kingdom, 2019.
[249] Rambla de Morales; Planta desaladora de agua de mar con capacidad para producir 60,000 m^3/día de agua apta para riego, Infoenviro (Infoenviro, 2006, June, 35–50).
[250] Trends & Developments in Advanced Energy Recovery Technologies for Desalination, Energy Recovery Inc. (ERI), US, 2013.
[251] Energy recovery device is designed to make desalination affordable, Membrane Technology, 2008 (2008) 9–10, https://doi.org/10.1016/S0958-2118(08)70064-8
[252] acuaMed, Torrevieja Desalination Plant, http://www.acuamed.es/media/actuaciones/109/torrevieja-corta-def.pdf (accessed September 2018).
[253] Annual Report 2013, acuaMed, Madrid, 2014.
[254] R. Franks, Analyzing three years of SWRO plant operation at elevated feed pH to save energy and improve boron rejection, in: International Desalination Association World Congress on Desalination and Water Reuse 2013, Tianjin, October, 2013.
[255] acuaMed, Valdelentisco Desalination Plant, http://www.acuamed.es/media/actuaciones/87/valdelentisco-corta-def.pdf (accessed September 2018).

[256] A. Belatoui, H. Bouabessalam, O. Rouane Hacene, J.A. De-la-Ossa-Carretero, E. Martinez-Garcia, J. Sanchez-Liza, Environmental effects of brine discharge from two desalination plants in Algeria (South Western Mediterranean), Desalination and Water Treatment, 76 (2017) 311–318, https://doi.org/10.5004/dwt.2017.20812

[257] E. Cardona, A. Piacentino, F. Marchese, Energy saving in two-stage reverse osmosis systems coupled with ultrafiltration processes, Desalination, 184 (2005) 125–137, https://doi.org/10.1016/j.desal.2005.03.063

[258] Desalination plants developed under Project Finance (P3 model), http://www.abengoawater.com/export/sites/abengoawater/resources/pdf/Folleto-plantas-desaladoras.pdf (accessed September 2018).

[259] Honaine Desalination Plant (Algeria), Infoenviro, 2012, April, 29–34.

[260] M. Samir, K. Ahmed, Assessment of the quality of water discharges from a desalination plant: case of Honaine station (Western Algeria), International Journal of Sciences: Basic and Applied Research, 39 (2018) 89–95, https://www.gssrr.org/index.php/JournalOfBasicAndApplied/article/view/9069

[261] Global Water Intelligence (GWI), Water Desalination Report, 2013, 49, 32.

[262] G. Gude, Renewable Energy Powered Desalination Handbook: Application and Thermodynamics, Elsevier Science, Oxford, United Kingdom, 2018.

[263] A.A. Bakr, W.A. Makled, New pretreatment media filtration for SWRO membranes of desalination plants, Desalination and Water Treatment, 55 (2015) 718–730, https://doi.org/10.1080/19443994.2014.944219

[264] B. Martinez, Jorf Lasfar: the Largest SWRO Desalination Plant in Morocco, in, presented in part at the International Conference on Desalination and Sustainability, Casablanca, March, 2012.

[265] S. Le Gallou, S. Bertrand, K.H. Madan, Al Dur SWRO plant: a double challenge for pre- and post-treatments, Desalination and Water Treatment, 51 (2013) 101–110, https://doi.org/10.1080/19443994.2012.694235

[266] SUEZ, Al Dur seawater reverse osmosis desalination plant, https://www.suezwaterhandbook.com/content/download/5689/91232/version/5/file/AL_DUR_EN_A4.pdf (accessed September 2018).

[267] M. Marzooq, M. Alsabbagh, W. Al-Zubari, Energy consumption in the municipal water supply sector in the kingdom of Bahrain, Computational Water, Energy, and Environmental Engineering, 7 (2018) 95–110, https://doi.org/10.4236/cweee.2018.73006

[268] G. Schneier-Madanes, M.-F. Courel, Water and sustainability in arid regions, High Demand in a Land of Water Scarcity: Iran, Zehtabian, G., H. Khosravi and M. Ghodsi (Eds.). Springer, London, New York, ISBN, 13 (2010) 75–86.

[269] B. Sauvet-Goichon, Ashkelon desalination plant – a successful challenge, Desalination, 203 (2007) 75–81, https://doi.org/10.1016/j.desal.2006.03.525

[270] A. Hermony, I. Sutzkover-Gutman, Y. Talmi, O. Fine, Palmachim Seawater desalination plant – seven years of expansions with uninterrupted operation together with process improvements, Desalination and Water Treatment, 55 (2015) 2526–2535, https://doi.org/10.1080/19443994.2014.940207

[271] Y. Hanegbi, Y. Mansdorf, Status of palmachim desalination project, in: 8th Annual Conference, Haifa, Israel, Israel Desalination Society, 2006, pp. 11–16.

[272] A. Yechiel, Y. Shevah, Optimization of energy costs for SWRO desalination plants, Desalination and Water Treatment, 46 (2012) 304–311, https://doi.org/10.1080/19443994.2012.677561

[273] A.H. Gulamhusein, A.S. Al Sheikh Khalil, I.A. Fatah, R. Boda, S. Rybar, IMS SWRO Kindasa – two years of operational experience, Desalination and Water Treatment, 10 (2009) 245–254, https://doi.org/10.5004/dwt.2009.823

[274] Pretreatment, AMTA, Winter 2011/2012 (2012).
[275] K. Tanaka, K. Matsui, T. Hori, H. Iwahashi, K. Takeuchi, Y. Ito, World's first large full triple-pass RO seawater desalination plant, Mitsubishi Heavy Industries Technical Review, 46 (2009) 12–14, https://www.mhi.co.jp/technology/review/pdf/e461/e461012.pdf.
[276] Global Water Intelligence Magazine, 11 (2010).
[277] V. Van Der Mast, T. Altmann, T. Attenborough, T. Hori, Y. Ito, H. Iwahashi, F. Mubeen, New Shuqaiq SWRO Desalination Plant with Advance BWRO Process, in: IDA World Congress 2007, Mas Palomas, October, 2007.
[278] A.D. Khawaji, I.K. Kutubkhanah, J.-M. Wie, A 13.3 MGD seawater RO desalination plant for Yanbu Industrial City, Desalination, 203 (2007) 176–188, https://doi.org/10.1016/j.desal.2006.02.018
[279] SUR Desalination plant, Sharqiyah Desalination Company, N/A.
[280] P. Choules, Update on large scale ocean desal plants, in: Ii Seminario Internacional De Desalacion, Antofagasta, 2012.
[281] A.D. Khawaji, I.K. Kutubkhanah, J.-M. Wie, Advances in seawater desalination technologies, Desalination, 221 (2008) 47–69, https://doi.org/10.1016/j.desal.2007.01.067
[282] M. Heberger, Australia's millennium drought: impacts and responses, in: P.H. Gleick (Ed.) The World's Water, Island Press, Washington DC, 2012, pp. 97–125. https://doi.org/10.1007/978-1-59726-228-6
[283] C. Harris, Energy recovery for membrane desalination, Desalination, 125 (1999) 173–180, https://doi.org/10.1016/S0011-9164(99)00136-8
[284] O.M. Al-Hawaj, The work exchanger for reverse osmosis plants, Desalination, 157 (2003) 23–27, https://doi.org/10.1016/S0011-9164(03)00378-3
[285] M.J. Guirguis, Energy Recovery Devices in Seawater Reverse Osmosis Desalination Plants with Emphasis on Efficiency and Economical Analysis of Isobaric Versus Centrifugal Devices, University of South Florida, US, 2011.
[286] M. Kurihara, H. Takeuchi, SWRO-PRO system in 'Mega-ton water system' for energy reduction and low environmental impact, Water, 10 (2018) 48, https://doi.org/10.3390/w10010048
[287] M. Nair, D. Kumar, Water desalination and challenges: The Middle East perspective: a review, Desalination and Water Treatment, 51 (2013) 2030–2040, https://doi.org/10.1080/19443994.2013.734483
[288] F. El-Gohary, A. Tawfik, U. Mahmoud, Comparative study between chemical coagulation/precipitation (C/P) versus coagulation/dissolved air flotation (C/DAF) for pre-treatment of personal care products (PCPs) wastewater, Desalination, 252 (2010) 106–112, https://doi.org/10.1016/j.desal.2009.10.016
[289] M. Busch, R. Chu, U. Kolbe, Q. Meng, S. Li, Ultrafiltration pretreatment to reverse osmosis for seawater desalination – three years field experience in the Wangtan Datang power plant, Desalination and Water Treatment, 10 (2009) 1–20, https://doi.org/10.5004/dwt.2009.702
[290] A. Alhadidi, A.J.B. Kemperman, R. Schurer, J.C. Schippers, M. Wessling, W.G.J. van der Meer, Using SDI, SDI+ and MFI to evaluate fouling in a UF/RO desalination pilot plant, Desalination, 285 (2012) 153–162, https://doi.org/10.1016/j.desal.2011.09.049
[291] P. Glueckstern, M. Priel, Comparative cost of UF vs conventional pretreatment for SWRO systems, International Desalination and Water Reuse Quarterly, 13 (2003) 34–39, https://www.researchgate.net/profile/M-Priel/publication/266499481_Comparative_Cost_of_UF_vs_Conventional_Pretreatment_for_SWRO_Systems/links/55016d500cf2de950a727571/Comparative-Cost-of-UF-vs-Conventional-Pretreatment-for-SWRO-Systems.pdf

[292] A. Sambidge, "Red tide" forces desalination plant closure. ArabianBusiness.com, (2008), http://www.arabianbusiness.com/538468-red-tide-forces-desalination-plant-closure

[293] T. Pankratz, Red tides close desal plants, Water Desalination Report, 44 (2008).

[294] A. Alkaisi, R. Mossad, A. Sharifian-Barforoush, A review of the water desalination systems integrated with renewable energy, Energy Procedia, 110 (2017) 268–274, https://doi.org/10.1016/j.egypro.2017.03.138

[295] N. Voutchkov, Energy use for membrane seawater desalination – current status and trends, Desalination, 431 (2018) 2–14, https://doi.org/10.1016/j.desal.2017.10.033

[296] N. Ghaffour, S. Lattemann, T. Missimer, K.C. Ng, S. Sinha, G. Amy, Renewable energy-driven innovative energy-efficient desalination technologies, Applied Energy, 136 (2014) 1155–1165, https://doi.org/10.1016/j.apenergy.2014.03.033

[297] N. Kishizawa, K. Tsuzuki, M. Hayatsu, Low pressure multi-stage RO system developed in 'Mega-ton water system' for large-scaled SWRO plant, Desalination, 368 (2015) 81–88, https://doi.org/10.1016/j.desal.2015.01.045

[298] K. Spiegler, Y. El-Sayed, The energetics of desalination processes, Desalination, 134 (2001) 109–128, https://doi.org/10.1016/S0011-9164(01)00121-7

[299] A. Blanco-Marigorta, A. Lozano-Medina, J. Marcos, A critical review of definitions for exergetic efficiency in reverse osmosis desalination plants, Energy, 137 (2017) 752–760, https://doi.org/10.1016/j.energy.2017.05.136

[300] L. Fitzsimons, B. Corcoran, P. Young, G. Foley, Exergy analysis of water purification and desalination: a study of exergy model approaches, Desalination, 359 (2015) 212–224, https://doi.org/10.1016/j.desal.2014.12.033

[301] M.H. Sharqawy, S.M. Zubair, J.H. Lienhard V, Second law analysis of reverse osmosis desalination plants: an alternative design using pressure retarded osmosis, Energy, 36 (2011) 6617–6626, https://doi.org/10.1016/j.energy.2011.08.056

[302] M.H. Sharqawy, S.M. Zubair, J.H. Lienhard V, On exergy calculations of seawater with applications in desalination systems, International Journal of Thermal Sciences, 50 (2011) 187–196, https://doi.org/10.1016/j.ijthermalsci.2010.09.013

[303] P.W. Atkins, J. De Paula, J. Keeler, Atkins' Physical Chemistry, Oxford University Press, Oxford, United Kingdom, 2018.

[304] J.M. Smith, H.C. Van Ness, M.M. Abbott, Introduction to Chemical Engineering Thermodynamics (8th ed.), McGraw-Hill, New York, 2017.

[305] R.K. McGovern, On the potential of forward osmosis to energetically outperform reverse osmosis desalination, Journal of Membrane Science, 469 (2014) 245–250, https://doi.org/10.1016/j.memsci.2014.05.061

[306] A. Ghobeity, A. Mitsos, Optimal time-dependent operation of seawater reverse osmosis, Desalination, 263 (2010) 76–88, https://doi.org/10.1016/j.desal.2010.06.041

[307] S. Loutatidou, N. Liosis, R. Pohl, T.B. Ouarda, H.A. Arafat, Wind-powered desalination for strategic water storage: techno-economic assessment of concept, Desalination, 408 (2017) 36–51, https://doi.org/10.1016/j.desal.2017.01.002

[308] K.S. Al-Malahy, T. Hodgkiess, Comparative studies of the seawater corrosion behaviour of a range of materials, Desalination, 158 (2003) 35–42, https://doi.org/10.1016/S0011-9164(03)00430-2

[309] E.M. Hoek, J. Allred, T. Knoell, B.-H. Jeong, Modeling the effects of fouling on full-scale reverse osmosis processes, Journal of Membrane Science, 314 (2008) 33–49, https://doi.org/10.1016/j.memsci.2008.01.025

[310] Y. Du, L. Xie, Y. Liu, S. Zhang, Y. Xu, Optimization of reverse osmosis networks with split partial second pass design, Desalination, 365 (2015) 365–380, https://doi.org/10.1016/j.desal.2015.03.019

[311] Y. Lu, A. Liao, Y. Hu, Design of reverse osmosis networks for multiple freshwater production, Korean Journal of Chemical Engineering, 30 (2013) 988–996, https://doi.org/10.1007/s11814-013-0009-8

[312] Y. Yang, Y. Sun, X. Song, J. Yu, Separation of mono- and di-valent ions from seawater reverse osmosis brine using selective electrodialysis, Environmental Science and Pollution Research, 28 (2021) 18754–18767, https://doi.org/10.1007/s11356-020-10014-9

[313] A.P. Straub, A. Deshmukh, M. Elimelech, Pressure-retarded osmosis for power generation from salinity gradients: is it viable?, Energy & Environmental Science, 9 (2016) 31–48, https://doi.org/10.1039/C5EE02985F

[314] S. Lin, A.P. Straub, M. Elimelech, Thermodynamic limits of extractable energy by pressure retarded osmosis, Energy Environmental Science, 7 (2014) 2706–2714, https://doi.org/10.1039/C4EE01020E

[315] M. Kurihara, H. Yamamura, T. Nakanishi, High recovery/high pressure membranes for brine conversion SWRO process development and its performance data, Desalination, 125 (1999) 9–15, https://doi.org/10.1016/S0011-9164(99)00119-8

[316] Y.-Y. Lu, Y.-D. Hu, X.-L. Zhang, L.-Y. Wu, Q.-Z. Liu, Optimum design of reverse osmosis system under different feed concentration and product specification, Journal of Membrane Science, 287 (2007) 219–229, https://doi.org/10.1016/j.memsci.2006.10.037

[317] Q.J. Wei, R.K. McGovern, Saving energy with an optimized two-stage reverse osmosis system, Environmental Science: Water Research & Technology, 3 (2017) 659–670, https://doi.org/10.1039/C7EW00069C

[318] T. Trifonova, L. Shirkin, O. Selivanov, M. Ilina, A. Povorov, Evaluating membrane separation effectiveness for highly concentrated aqueous media applying two-stage reverse osmosis plant, International Journal of Emerging Trends in Engineering Research, 8 (2020) 906–910, https://doi.org/10.30534/ijeter/2020/47832020

[319] M. Al-Obaidi, C. Kara-Zaitri, I.M. Mujtaba, Simulation and optimisation of a two-stage/two-pass reverse osmosis system for improved removal of chlorophenol from wastewater, Journal of Water Process Engineering, 22 (2018) 131–137, https://doi.org/10.1016/j.jwpe.2018.01.012

[320] A. Efraty, in, US Pat., 7.695,614 B2, 2010.

[321] R.L. Stover, High recovery, low fouling, and low energy reverse osmosis, Desalination and Water Treatment, 57 (2016) 26501–26506, https://doi.org/10.1080/19443994.2016.1168586

[322] J.R. Werber, A. Deshmukh, M. Elimelech, Can batch or semi-batch processes save energy in reverse-osmosis desalination?, Desalination, 402 (2017) 109–122, https://doi.org/10.1016/j.desal.2016.09.028

[323] D.M. Warsinger, E.W. Tow, K.G. Nayar, L.A. Maswadeh, Energy efficiency of batch and semi-batch (CCRO) reverse osmosis desalination, Water Research, 106 (2016) 272–282, https://doi.org/10.1016/j.watres.2016.09.029

[324] A. Efraty, Z. Gal, Closed circuit desalination series No 7: retrofit design for improved performance of conventional BWRO system, Desalination and Water Treatment, 41 (2012) 301–307, https://doi.org/10.1080/19443994.2012.664716

[325] B. Sutariya, H. Raval, Analytical study of optimum operating conditions in semi-batch closed-circuit reverse osmosis (CCRO), Separation and Purification Technology, 264 (2021) 118421, https://doi.org/10.1016/j.seppur.2021.118421

[326] T. Lee, A. Rahardianto, Y. Cohen, Multi-cycle operation of semi-batch reverse osmosis (SBRO) desalination, Journal of Membrane Science, 588 (2019) 117090, https://doi.org/10.1016/j.memsci.2019.05.015

[327] S. Li, K. Duran, S. Delagah, J. Mouawad, X. Jia, M. Sharbatmaleki, Energy efficiency of staged reverse osmosis (RO) and closed-circuit reverse osmosis

(CCRO) desalination: a model-based comparison, Water Supply, 20 (2020) 3096–3106, https://doi.org/10.2166/ws.2020.208

[328] D.M. Warsinger, E.W. Tow, K.G. Nayar, L.A. Maswadeh, J.H. Lienhard, Energy efficiency of batch and semi-batch (CCRO) reverse osmosis desalination, Water Research, 106 (2016) 272–282, https://doi.org/10.1016/j.watres.2016.09.029

[329] M. Barello, D. Manca, R. Patel, I.M. Mujtaba, Operation and modeling of RO desalination process in batch mode, Computers & Chemical Engineering, 83 (2015) 139–156, https://doi.org/10.1016/j.compchemeng.2015.05.022

[330] T. Qiu, P.A. Davies, Longitudinal dispersion in spiral wound RO modules and its effect on the performance of batch mode RO operations, Desalination, 288 (2012) 1–7, https://doi.org/10.1016/j.desal.2011.11.054

[331] T. Qiu, P.A. Davies, Comparison of configurations for high-recovery inland desalination systems, Water, 4 (2012) 690–706, https://doi.org/10.3390/w4030690

[332] T. Qiu, P. Davies, Concentration polarization model of spiral-wound membrane modules with application to batch-mode RO desalination of brackish water, Desalination, 368 (2015) 36–47, https://doi.org/10.1016/j.desal.2014.12.048

[333] P.A. Davies, J. Wayman, C. Alatta, K. Nguyen, J. Orfi, A desalination system with efficiency approaching the theoretical limits, Desalination and Water Treatment, 57 (2016) 23206–23216, https://doi.org/10.1080/19443994.2016.1180482

[334] D.M. Warsinger, E.W. Tow, L.A. Maswadeh, G.B. Connors, J. Swaminathan, J.H. Lienhard V, Inorganic fouling mitigation by salinity cycling in batch reverse osmosis, Water Research, 137 (2018) 384–394, https://doi.org/10.1016/j.watres.2018.01.060

[335] J.H. Lienhard, G.P. Thiel, D.M. Warsinger, L.D. Banchik, Low carbon desalination: status and research, development, and demonstration needs, Report of a Workshop Conducted at the Massachusetts Institute of Technology in Association with the Global Clean Water Desalination Alliance (2016).

[336] M. Li, Dynamic operation of batch reverse osmosis and batch pressure retarded osmosis, Industrial & Engineering Chemistry Research, 59 (2020) 3097–3108, https://doi.org/10.1021/acs.iecr.9b06371

[337] M. Li, Y. Heng, J. Luo, Batch reverse osmosis: a new research direction in water desalination, Science Bulletin, 65 (2020) 1705–1708, https://doi.org/10.1016/j.scib.2020.05.032

[338] Q.J. Wei, C.I. Tucker, P.J. Wu, A.M. Trueworthy, E.W. Tow, Impact of salt retention on true batch reverse osmosis energy consumption: experiments and model validation, Desalination, 479 (2020) 114177, https://doi.org/10.1016/j.desal.2019.114177

[339] J. Swaminathan, E.W. Tow, R.L. Stover, Practical aspects of batch RO design for energy-efficient seawater desalination, Desalination, 470 (2019) 114097, https://doi.org/10.1016/j.desal.2019.114097

[340] A. Chougradi, F. Zaviska, A. Abed, J. Harmand, J.-E. Jellal, M. Heran, Batch reverse osmosis desalination modeling under a time-dependent pressure profile, Membranes, 11 (2021) 173, https://doi.org/10.3390/membranes11030173

[341] H. Abu Ali, M. Baronian, L. Burlace, P.A. Davies, S. Halasah, M. Hind, A. Hossain, C. Lipchin, A. Majali, M. Mark, Off-grid desalination for irrigation in the Jordan Valley, Desalination and Water Treatment, 168 (2019) 143–154, https://doi.org/10.5004/dwt.2019.24567

[342] A. Das, D.M. Warsinger, Batch counterflow reverse osmosis, Desalination, 507 (2021) 115008, https://doi.org/10.1016/j.desal.2021.115008

[343] S. Cordoba, A. Das, J. Leon, J.M. Garcia, D.M. Warsinger, Double-acting batch reverse osmosis configuration for best-in-class efficiency and low downtime, Desalination, 506 (2021) 114959, https://doi.org/10.1016/j.desal.2021.114959

[344] P. Davies, A. Afifi, F. Khatoon, G. Kuldip, S. Javed, S. Khan, Double-acting batch-RO system for desalination of brackish water with high efficiency and high recovery, Desalination and Water Treatment, 224 (2021) 1–11, https://doi.org/10.5004/dwt.2021.26995

[345] Q.J. Wei, C.I. Tucker, P.J. Wu, A.M. Trueworthy, E.W. Tow, J.H. Lienhard V, True batch reverse osmosis prototype: model validation and energy savings, in: The International Desalination Association World Congress on Desalination and Water Reuse 2019, International Desalination Association, Dubai, UAE, 2019.

[346] Q.J. Wei, C.I. Tucker, P.J. Wu, A.M. Trueworthy, E.W. Tow, J.H. Lienhard V, Impact of salt retention on true batch reverse osmosis energy consumption: experiments and model validation, Desalination, 479 (2020) 114177, https://doi.org/10.1016/j.desal.2019.114177

[347] Q.J. Wei, C.I. Tucker, P.J. Wu, A.M. Trueworthy, E.W. Tow, J.H. Lienhard V, Batch reverse osmosis: experimental results, model validation, and design implications, in: 2019 AMTA/AWWA Membrane Technology Conference & Exposition, American Membrane Technology Association, New Orleans, Louisiana, 2019.

[348] D.M. Warsinger, E.W. Tow, L.A. Maswadeh, G.B. Connors, J. Swaminathan, Inorganic fouling mitigation by salinity cycling in batch reverse osmosis, Water Research, 137 (2018) 384–394, https://doi.org/10.1016/j.watres.2018.01.060

[349] E. Jones, M. Qadir, M.T. van Vliet, V. Smakhtin, S.-M. Kang, The state of desalination and brine production: a global outlook, Science of the Total Environment, 657 (2019) 1343–1356, https://doi.org/10.1016/j.scitotenv.2018.12.076

[350] T. Tong, M. Elimelech, The global rise of zero liquid discharge for wastewater management: drivers, technologies, and future directions, Environmental Science & Technology, 50 (2016) 6846–6855, https://doi.org/10.1021/acs.est.6b01000

[351] K.J. Lu, Z.L. Cheng, J. Chang, L. Luo, T.-S. Chung, Design of zero liquid discharge desalination (ZLDD) systems consisting of freeze desalination, membrane distillation, and crystallization powered by green energies, Desalination, 458 (2019) 66–75, https://doi.org/10.1016/j.desal.2019.02.001

[352] R. Schwantes, K. Chavan, D. Winter, C. Felsmann, J. Pfafferott, Techno-economic comparison of membrane distillation and MVC in a zero liquid discharge application, Desalination, 428 (2018) 50–68, https://doi.org/10.1016/j.desal.2017.11.026

[353] Z. Wang, A. Deshmukh, Y. Du, M. Elimelech, Minimal and zero liquid discharge with reverse osmosis using low-salt-rejection membranes, Water Research, 170 (2020) 115317, https://doi.org/10.1016/j.watres.2019.115317

[354] Q.J. Wei, R.K. McGovern, J.H. Lienhard V, Saving energy with an optimized two-stage reverse osmosis system, Environmental Science: Water Research & Technology, 3 (2017) 659–670, https://doi.org/10.1039/C7EW00069C

[355] T.V. Bartholomew, L. Mey, J.T. Arena, N.S. Siefert, M.S. Mauter, Osmotically assisted reverse osmosis for high salinity brine treatment, Desalination, 421 (2017) 3–11, https://doi.org/10.1016/j.desal.2017.04.012

[356] K.H. Chu, J. Lim, S.-J. Kim, T.-U. Jeong, M.-H. Hwang, Determination of optimal design factors and operating conditions in a large-scale seawater reverse osmosis desalination plant, Journal of Cleaner Production, 244 (2020) 118918, https://doi.org/10.1016/j.jclepro.2019.118918

[357] S. Haryati, A.B. Hamzah, P.S. Goh, M.S. Abdullah, A.F. Ismail, M.D. Bustan, Process intensification of seawater reverse osmosis through enhanced train capacity and module size – simulation on Lanzarote IV SWRO plant, Desalination, 408 (2017) 92–101, https://doi.org/10.1016/j.desal.2017.01.011

[358] K. Jeong, M. Park, S.J. Ki, J.H. Kim, A systematic optimization of internally staged design (ISD) for a full-scale reverse osmosis process, Journal of Membrane Science, 540 (2017) 285–296, https://doi.org/10.1016/j.memsci.2017.06.066

[359] S.-J. Im, J. Choi, S. Jeong, A. Jang, New concept of pump-less forward osmosis (FO) and low-pressure membrane (LPM) process, Scientific Reports, 7 (2017) 14569, https://doi.org/10.1038/s41598-017-15274-z

[360] T.Y. Cath, N.T. Hancock, C.D. Lundin, C. Hoppe-Jones, J.E. Drewes, A multi-barrier osmotic dilution process for simultaneous desalination and purification of impaired water, Journal of Membrane Science, 362 (2010) 417–426, https://doi.org/10.1016/j.memsci.2010.06.056

[361] B.G. Choi, M. Zhan, K. Shin, S. Lee, S. Hong, Pilot-scale evaluation of FO-RO osmotic dilution process for treating wastewater from coal-fired power plant integrated with seawater desalination, Journal of Membrane Science, 540 (2017) 78–87, https://doi.org/10.1016/j.memsci.2017.06.036

[362] M. Zhan, Y. Kim, J. Lim, S. Hong, Application of fouling index for forward osmosis hybrid system: a pilot demonstration, Journal of Membrane Science, 617 (2021) 118624, https://doi.org/10.1016/j.memsci.2020.118624

[363] D.I. Kim, J. Kim, H.K. Shon, S. Hong, Pressure retarded osmosis (PRO) for integrating seawater desalination and wastewater reclamation: energy consumption and fouling, Journal of Membrane Science, 483 (2015) 34–41, https://doi.org/10.1016/j.memsci.2015.02.025

[364] R.V. Linares, Z. Li, V. Yangali-Quintanilla, N. Ghaffour, G. Amy, T. Leiknes, J. Vrouwenvelder, Life cycle cost of a hybrid forward osmosis–low pressure reverse osmosis system for seawater desalination and wastewater recovery, Water Research, 88 (2016) 225–234, https://doi.org/10.1016/j.watres.2015.10.017

[365] V. Yangali-Quintanilla, Z. Li, R. Valladares, Q. Li, G. Amy, Indirect desalination of Red Sea water with forward osmosis and low pressure reverse osmosis for water reuse, Desalination, 280 (2011) 160–166, https://doi.org/10.1016/j.desal.2011.06.066

[366] J. Kim, G. Blandin, S. Phuntsho, A. Verliefde, P. Le-Clech, H. Shon, Practical considerations for operability of an 8″ spiral wound forward osmosis module: hydrodynamics, fouling behaviour and cleaning strategy, Desalination, 404 (2017) 249–258, https://doi.org/10.1016/j.desal.2016.11.004

[367] S. Loeb, Method and Apparatus for Generating Power Utilizing Pressure-Retarded Osmosis, Patent in United States, 1980

[368] A. Achilli, A.E. Childress, Pressure retarded osmosis: from the vision of Sidney Loeb to the first prototype installation – review, Desalination, 261 (2010) 205–211, https://doi.org/10.1016/j.desal.2010.06.017

[369] N.Y. Yip, A. Tiraferri, W.A. Phillip, J.D. Schiffman, L.A. Hoover, Y.C. Kim, M. Elimelech, Thin-film composite pressure retarded osmosis membranes for sustainable power generation from salinity gradients, Environmental Science & Technology, 45 (2011) 4360–4369, https://doi.org/10.1021/es104325z

[370] K.L. Lee, R.W. Baker, H.K. Lonsdale, Membranes for power generation by pressure-retarded osmosis, Journal of Membrane Science, 8 (1981) 141–171, https://doi.org/10.1016/S0376-7388(00)82088-8

[371] A. Tanioka, M. Kurihara, H. Sakai, Chapter 12 – megaton water system: high salinity pressure retarded osmosis, in: S. Sarp, N. Hilal (Eds.) Membrane-Based Salinity Gradient Processes for Water Treatment and Power Generation, Elsevier, 2018, pp. 319–334.

[372] J. Kim, M. Park, S.A. Snyder, J.H. Kim, Reverse osmosis (RO) and pressure retarded osmosis (PRO) hybrid processes: model-based scenario study, Desalination, 322 (2013) 121–130, https://doi.org/10.1016/j.desal.2013.05.010

[373] A. Achilli, J.L. Prante, N.T. Hancock, E.B. Maxwell, A.E. Childress, Experimental results from RO-PRO: a next generation system for low-energy desalination, Environmental Science & Technology, 48 (2014) 6437–6443, https://doi.org/10.1021/es405556s

[374] S. Lee, S.-H. Kim, Y.-G. Park, Chapter 11 – high-salinity pressure retarded osmosis using seawater reverse osmosis brine, in: S. Sarp, N. Hilal (Eds.) Membrane-Based Salinity Gradient Processes for Water Treatment and Power Generation, Elsevier, Amsterdam, Netherlands, 2018, pp. 297–316.
[375] S. Lee, T.-S. Park, Y.-G. Park, W.-I. Lee, S.-H. Kim, Toward scale-up of seawater reverse osmosis (SWRO) – pressure retarded osmosis (PRO) hybrid system: a case study of a 240 m^3/day pilot plant, Desalination, 491 (2020) 114429, https://doi.org/10.1016/j.desal.2020.114429
[376] K. Touati, H.S. Usman, C.N. Mulligan, M.S. Rahaman, Energetic and economic feasibility of a combined membrane-based process for sustainable water and energy systems, Applied Energy, 264 (2020) 114699, https://doi.org/10.1016/j.apenergy.2020.114699
[377] R.E. Pattle, Production of electric power by mixing fresh and salt water in the hydroelectric pile, Nature, 174 (1954) 660–660, https://doi.org/10.1038/174660a0
[378] P. Długołęcki, A. Gambier, K. Nijmeijer, M. Wessling, Practical potential of reverse electrodialysis as process for sustainable energy generation, Environmental Science & Technology, 43 (2009) 6888–6894, https://doi.org/10.1021/es9009635
[379] H. Tian, Y. Wang, Y. Pei, J.C. Crittenden, Unique applications and improvements of reverse electrodialysis: a review and outlook, Applied Energy, 262 (2020) 114482, https://doi.org/10.1016/j.apenergy.2019.114482
[380] N.Y. Yip, D.A. Vermaas, K. Nijmeijer, M. Elimelech, Thermodynamic, energy efficiency, and power density analysis of reverse electrodialysis power generation with natural salinity gradients, Environmental Science & Technology, 48 (2014) 4925–4936, https://doi.org/10.1021/es5005413
[381] D.A. Vermaas, J. Veerman, N.Y. Yip, M. Elimelech, M. Saakes, K. Nijmeijer, High efficiency in energy generation from salinity gradients with reverse electrodialysis, ACS Sustainable Chemistry & Engineering, 1 (2013) 1295–1302, https://doi.org/10.1021/sc400150w
[382] D.A. Vermaas, M. Saakes, K. Nijmeijer, Doubled power density from salinity gradients at reduced intermembrane distance, Environmental Science & Technology, 45 (2011) 7089–7095, https://doi.org/10.1021/es2012758
[383] W. Li, W.B. Krantz, E.R. Cornelissen, J.W. Post, A.R.D. Verliefde, C.Y. Tang, A novel hybrid process of reverse electrodialysis and reverse osmosis for low energy seawater desalination and brine management, Applied Energy, 104 (2013) 592–602, https://doi.org/10.1016/j.apenergy.2012.11.064
[384] E. Farrell, M.I. Hassan, R.A. Tufa, A. Tuomiranta, A.H. Avci, A. Politano, E. Curcio, H.A. Arafat, Reverse electrodialysis powered greenhouse concept for water- and energy-self-sufficient agriculture, Applied Energy, 187 (2017) 390–409, https://doi.org/10.1016/j.apenergy.2016.11.069
[385] Ramato A. Tufa, E. Curcio, E. Brauns, W. van Baak, E. Fontananova, G. Di Profio, Membrane distillation and reverse electrodialysis for near-zero liquid discharge and low energy seawater desalination, Journal of Membrane Science, 496 (2015) 325–333, https://doi.org/10.1016/j.memsci.2015.09.008
[386] R.A. Tufa, Y. Noviello, G. Di Profio, F. Macedonio, A. Ali, E. Drioli, E. Fontananova, K. Bouzek, E. Curcio, Integrated membrane distillation-reverse electrodialysis system for energy-efficient seawater desalination, Applied Energy, 253 (2019) 113551, https://doi.org/10.1016/j.apenergy.2019.113551
[387] J. Choi, Y. Oh, S. Chae, S. Hong, Membrane capacitive deionization-reverse electrodialysis hybrid system for improving energy efficiency of reverse osmosis seawater desalination, Desalination, 462 (2019) 19–28, https://doi.org/10.1016/j.desal.2019.04.003

[388] M.A. Anderson, A.L. Cudero, J. Palma, Capacitive deionization as an electrochemical means of saving energy and delivering clean water. Comparison to present desalination practices: will it compete?, Electrochimica Acta, 55 (2010) 3845–3856, https://doi.org/10.1016/j.electacta.2010.02.012
[389] N.Y. Yip, M. Elimelech, Comparison of energy efficiency and power density in pressure retarded osmosis and reverse electrodialysis, Environmental Science & Technology, 48 (2014) 11002–11012, https://doi.org/10.1021/es5029316
[390] N.Y. Yip, M. Elimelech, Influence of natural organic matter fouling and osmotic backwash on pressure retarded osmosis energy production from natural salinity gradients, Environmental Science & Technology, 47 (2013) 12607–12616, https://doi.org/10.1021/es403207m
[391] D.A. Vermaas, D. Kunteng, M. Saakes, K. Nijmeijer, Fouling in reverse electrodialysis under natural conditions, Water Research, 47 (2013) 1289–1298, https://doi.org/10.1016/j.watres.2012.11.053
[392] W. Suwaileh, D. Johnson, N. Hilal, Membrane desalination and water re-use for agriculture: state of the art and future outlook, Desalination, 491 (2020) 114559, https://doi.org/10.1016/j.desal.2020.114559
[393] S. Hong, I.-C. Kim, T. Tak, Y.-N. Kwon, Interfacially synthesized chlorine-resistant polyimide thin film composite (TFC) reverse osmosis (RO) membranes, Desalination, 309 (2013) 18–26, https://doi.org/10.1016/j.desal.2012.09.025
[394] S. Yu, M. Liu, Z. Lü, Y. Zhou, C. Gao, Aromatic-cycloaliphatic polyamide thin-film composite membrane with improved chlorine resistance prepared from m-phenylenediamine-4-methyl and cyclohexane-1, 3, 5-tricarbonyl chloride, Journal of Membrane Science, 344 (2009) 155–164, https://doi.org/10.1016/j.memsci.2009.07.046
[395] A.K. Ghosh, B.-H. Jeong, X. Huang, E.M.V. Hoek, Impacts of reaction and curing conditions on polyamide composite reverse osmosis membrane properties, Journal of Membrane Science, 311 (2008) 34–45, https://doi.org/10.1016/j.memsci.2007.11.038
[396] M.A. Kuehne, R.Q. Song, N.N. Li, R. J. Petersen, Flux enhancement in TFC RO membranes, Environmental Progress, 20 (2001) 23–26, https://doi.org/10.1002/ep.670200112
[397] D. Mukherjee, A. Kulkarni, W.N. Gill, Chemical treatment for improved performance of reverse osmosis membranes, Desalination, 104 (1996) 239–249, https://doi.org/10.1016/0011-9164(96)00047-1
[398] M. Elimelech, X. Zhu, A. E. Childress, S. Hong, Role of membrane surface morphology in colloidal fouling of cellulose acetate and composite aromatic polyamide reverse osmosis membranes, Journal of Membrane Science, 127 (1997) 101–109, https://doi.org/10.1016/S0376-7388(96)00351-1
[399] W. Choi, C. Lee, D. Lee, Y.J. Won, G.W. Lee, M.G. Shin, B. Chun, T.-S. Kim, H.-D. Park, H.W. Jung, J. S. Lee, J.-H. Lee, Sharkskin-mimetic desalination membranes with ultralow biofouling, Journal of Materials Chemistry A, 6 (2018) 23034–23045, https://doi.org/10.1039/C8TA06125D
[400] O. Heinz, M. Aghajani, A. R. Greenberg, Y. Ding, Surface-patterning of polymeric membranes: fabrication and performance, Current Opinion in Chemical Engineering, 20 (2018) 1–12, https://doi.org/10.1016/j.coche.2018.01.008
[401] C. Tang, Z. Wang, I. Petrinić, A. G. Fane, C. Hélix-Nielsen, Biomimetic aquaporin membranes coming of age, Desalination, 368 (2015) 89–105, https://doi.org/10.1016/j.desal.2015.04.026
[402] Y. Zhao, C. Qiu, X. Li, A. Vararattanavech, W. Shen, J. Torres, C. Helix-Nielsen, R. Wang, X. Hu, A.G. Fane, C.Y. Tang, Synthesis of robust and high-performance

aquaporin-based biomimetic membranes by interfacial polymerization-membrane preparation and RO performance characterization, Journal of Membrane Science, 423 (2012) 422–428, https://doi.org/10.1016/j.memsci.2012.08.039

[403] D. Wu, X. Liu, S. Yu, M. Liu, C. Gao, Modification of aromatic polyamide thin-film composite reverse osmosis membranes by surface coating of thermo-responsive copolymers P (NIPAM-co-Am). I: preparation and characterization, Journal of Membrane Science, 352 (2010) 76–85, https://doi.org/10.1016/j.memsci.2010.01.061

[404] X. Zhu, D. Jassby, Electroactive membranes for water treatment: enhanced treatment functionalities, energy considerations, and future challenges, Accounts of Chemical Research, 52 (2019) 1177–1186, https://doi.org/10.1021/acs.accounts.8b00558

[405] M. Majumder, N. Chopra, B.J. Hinds, Mass transport through carbon nanotube membranes in three different regimes: ionic diffusion and gas and liquid flow, ACS Nano, 5 (2011) 3867–3877, https://doi.org/10.1021/nn200222g

[406] J.K. Holt, H.G. Park, Y. Wang, M. Stadermann, A.B. Artyukhin, C.P. Grigoropoulos, A. Noy, O. Bakajin, Fast mass transport through sub-2-nanometer carbon nanotubes, Science, 312 (2006) 1034–1037, https://doi.org/10.1126/science.1126298

[407] E. Yang, A.B. Alayande, K. Goh, C.-M. Kim, K.-H. Chu, M.-H. Hwang, J.-H. Ahn, K.-J. Chae, 2D materials-based membranes for hydrogen purification: current status and future prospects, International Journal of Hydrogen Energy, 46 (2020) 11389–11410, https://doi.org/10.1016/j.ijhydene.2020.04.053

[408] C. Siskens, Applications of ceramic membranes in liquid filtration, in: Membrane Science and Technology, Elsevier, Amsterdam, Netherlands, 1996, pp. 619–639.

[409] L. Lia, J. Dong, T.M. Nenoff, R. Lee, Reverse osmosis of ionic aqueous solutions on aMFI zeolite membrane, Desalination, 170 (2004) 309–316, https://doi.org/10.1016/j.desal.2004.02.102

[410] N. Liu, L. Li, B. McPherson, R. Lee, Removal of organics from produced water by reverse osmosis using MFI-type zeolite membranes, Journal of Membrane Science, 325 (2008) 357–361, https://doi.org/10.1016/j.memsci.2008.07.056

[411] M.C. Duke, J. O'Brien-Abraham, N. Milne, B. Zhu, J.Y. Lin, J.C.D. da Costa, Seawater desalination performance of MFI type membranes made by secondary growth, Separation and Purification Technology, 68 (2009) 343–350, https://doi.org/10.1016/j.seppur.2009.06.003

[412] M.L. Lind, A.K. Ghosh, A. Jawor, X. Huang, W. Hou, Y. Yang, E.M. Hoek, Influence of zeolite crystal size on zeolite-polyamide thin film nanocomposite membranes, Langmuir, 25 (2009) 10139–10145, https://doi.org/10.1021/la900938x

[413] E. Okumus, T. Gurkan, L. Yilmaz, Development of a mixed-matrix membrane for pervaporation, Separation Science and Technology, 29 (1994) 2451–2473, https://doi.org/10.1080/01496399408002203

[414] A. Ismail, P. Goh, S. Sanip, M. Aziz, Transport and separation properties of carbon nanotube-mixed matrix membrane, Separation and Purification Technology, 70 (2009) 12–26, https://doi.org/10.1016/j.seppur.2009.09.002